Springer Complexity

Springer Complexity is an interdisciplinary program publishing the best research and academic-level teaching on both fundamental and applied aspects of complex systems—cutting across all traditional disciplines of the natural and life sciences, engineering, economics, medicine, neuroscience, social and computer science.

Complex Systems are systems that comprise many interacting parts with the ability to generate a new quality of macroscopic collective behavior the manifestations of which are the spontaneous formation of distinctive temporal, spatial or functional structures. Models of such systems can be successfully mapped onto quite diverse "real-life" situations like the climate, the coherent emission of light from lasers, chemical reaction-diffusion systems, biological cellular networks, the dynamics of stock markets and of the internet, earthquake statistics and prediction, freeway traffic, the human brain, or the formation of opinions in social systems, to name just some of the popular applications.

Although their scope and methodologies overlap somewhat, one can distinguish the following main concepts and tools: self-organization, nonlinear dynamics, synergetics, turbulence, dynamical systems, catastrophes, instabilities, stochastic processes, chaos, graphs and networks, cellular automata, adaptive systems, genetic algorithms and computational intelligence.

The three major book publication platforms of the Springer Complexity program are the monograph series "Understanding Complex Systems" focusing on the various applications of complexity, the "Springer Series in Synergetics", which is devoted to the quantitative theoretical and methodological foundations, and the "Springer Briefs in Complexity" which are concise and topical working reports, case studies, surveys, essays and lecture notes of relevance to the field. In addition to the books in these two core series, the program also incorporates individual titles ranging from textbooks to major reference works.

Indexed by SCOPUS, INSPEC, zbMATH, SCImago.

Understanding Complex Systems

Juan Guillermo Diaz Ochoa

Complexity Measurements and Causation for Dynamic Complex Systems

 Springer

Juan Guillermo Diaz Ochoa
PerMediQ
Stuttgart, Baden-Württemberg, Germany

ISSN 1860-0832   ISSN 1860-0840 (electronic)
Understanding Complex Systems
ISBN 978-3-031-84708-0   ISBN 978-3-031-84709-7 (eBook)
https://doi.org/10.1007/978-3-031-84709-7

This Springer imprint is published by the registered company Springer Nature Switzerland AG
The registered company address is: Gewerbestrasse 11, 6330 Cham, Switzerland

If disposing of this product, please recycle the paper.

This book is dedicated to the wind and to infinity, the birds and the mountains. And all those whom I deeply love and who have inspired me to try to better understand nature and the world.

Preface and Acknowledgments

This book is the result of my research exploring concepts of causality in complex multiscale systems, particularly considering the effect of context on the definition of microscopic states, as well as the need to coherently write the background necessary to understand a measure of complexity capable of capturing both the variability and autonomy of the system. This concept has implications for the understanding and subsequent control of any complex system and involves not only the technical but also the philosophical aspects that are addressed in this book.

However, there are also personal reasons that motivated me to write this book: my desire to understand nature, the observations and experiences hiking in the high mountains in the Andes (specifically the unique ecosystem of the Colombian Paramos), the Swabian Alb and the Alps and all their local cultures (some are lost forever), as well as my long years of research in academia and biomedicine.

This book is not only the product of curiosity but also the preoccupation of observing how humanity had displaced itself from its natural environment.

Many fundamental problems arise not only in the laboratory but also in the real world. I have also learned that a thorough understanding of the relationship between the "small" and the "large" is imperative. After all these scales are explored, one realizes not only how beautiful but also how fragile and powerful nature is and how important it is to carefully listen to what nature wants to communicate to us.

I would first like to thank my father and late mother for all the insightful conversations we had in my childhood and to both my wife and son for their constant support. I am deeply grateful to Prof. K. Binder, who has inspired me in statistical physics and was a valuable colleague exchanging ideas and thoughts.

The concept of elastic states, and the motivation behind the book, could not be possible without Elena Ramirez, our continuous lively conversation on behavioral economics, as well as our collaboration. Bernardo Diaz-Ramirez played an important role in discussions about the connection between mythology, story and science.

Many colleagues in the past (in my different stations) and students have also inspired my work and kept me motivated; for this I want, in particular, to express my gratitude to Anna Nitschke, who gave me the final push to embark on this book after our long personal, philosophical and technical discussion on complexity and machine learning in medicine—as well as feedback on the first proposal of this project—and Faizan E Mustafa, who has been an excellent colleague and partner to discuss several issues in philosophy and machine learning.

Additionally, special thanks to Martin Grundman for the exchange of ideas and technical background and Steffan Gryglewski for the constant exchange of physics and philosophy. Finally, I would like to thank June Othman for critical proofreading and her support, encouragement and contribution in making all these ideas clearer.

This book was written by a human without generative AI. Some proofreading has been performed via AI Tools[1]; however, the final proofs and corrections have been performed by humans. Thus, all the ideas and errors in this text are purely of human nature.

This book begins with a labyrinth, as described by Jorge Luis Borges:

No habrá nunca una puerta. Estás adentro
y el alcázar abarca el universo
y no tiene ni anverso ni reverso
ni externo muro ni secreto centro.

No esperes que el rigor de tu camino
que tercamente se bifurca en otro,
tendrá fin. Es de hierro tu destino como tu juez.

No aguardes la embestida
del toro que es un hombre y cuya extraña
forma plural da horror a la maraña
de interminable piedra entretejida.

No existe. Nada esperes. Ni siquiera
en el negro crepúsculo la fiera.

J. L. Borges

Stuttgart, Germany   Juan Guillermo Diaz Ochoa

---

[1] Wordtune and Curie.

Contents

Abbreviations and Mathematical Symbols

Abbreviations

AI	Artificial Intelligence
DL	Deep Learning
ECG	Electrocardiogram
EHR	Electronic Health Record
EMA	European medicine administration
FDA	Federal drug administration
GNN	Graph Neural Network
LLM	Large Language Models
LSTM	Large Short-Term Memory
MC	Monte Carlo
MEM	Modulus of Elasticity of Mechanisms
ML	Machine Learning
NLP	Natural Language Processing
OEE	Open Ended Evolution
SVM	Support Vector Machine
TDA	Topological Data Analysis

Mathematical Symbols

Γ	Trajectory
$\mathcal{P}(\widehat{M})$	Computation based on the internal structure $\widehat{M}$
$\widehat{M}$	Internal structure to store system's information
$\varkappa$	Control function
$\mathcal{R}_i$	Space
s_i	Interacting element i
σ_i	Observed state of interacting element i

ε	Environment of an element, explored with its trajectory Γ
g_{ij}	Connectivity factor
δ_i	Sequence of time (should not be confused with the Dirac's Delta)
w^λ	Point cloud (topological analysis)
K_B	Boltzmann Constant
T	Temperature
Z	Partition function
$P_{eq}(\{\sigma_i\})$	Probability function (in equilibrium)
O	Observable
J	Coupling constant in a lattice
$W_{i \rightarrow i'}$	Transition probability (for Monte Carlo methods)

Chapter 1
Introduction

*Omnes enim causae effectuum naturaliumdantur per lineas,
angulos et figuras. Aliter enim impossibile est scire propter quid
in illis, said one of the grand masters at Oxford. However, there
is still so much out there that is left unknown. So far, we have
only learned to orientate ourselves in the labyrinth. Now, we
have to dive deep to find out.*
U. Eco/Die Name der Rose

The complexity of the world is comparable to that of a labyrinth since its structure cannot be reduced to a simple path. Instead, the winding and connecting paths take up the entire interior and must be 'traversed' to obtain an idea of its architecture. Humanity attempts to escape (and/or maneuver in) this labyrinth called "complexity" by trying to understand and/or define its structure. However, this is a challenging task. Despite this, many scholars and scientists have not only claimed to have escaped this labyrinth but also postulated how to decipher every other labyrinth in the universe by assuming that all labyrinths are structured according to a similar universal pattern.

Labyrinths are a recurring theme in mythology, art, architecture (for example, in churches, they were used for penitential exercises and as a symbol of the search of God; see Fig. 1.1), and literature.

For example, in the novel 'In the Name of the Rose', written by Umberto Eco, a series of murders in a monastery in northern Italy describes as such and how William of Baskerville, a Franciscan monk with his disciple Adson, tried to solve the case. However, this book is not merely a crime story such as the medieval parody of Sherlock Holmes and Mr. Watson. Instead, it is an intellectual mystery that combines semiotics in fiction, biblical analysis, medieval studies, and literary theories [3]. The story is set in the monastery's library, containing a labyrinth where classical works were considered dangerous and were kept hidden. Finding the right way into the library is also a challenge for the protagonist and his disciple.

J. G. Diaz Ochoa, *Complexity Measurements and Causation for Dynamic Complex Systems*,
Understanding Complex Systems, https://doi.org/10.1007/978-3-031-84709-7_1

Fig. 1.1 In art and architecture, labyrinths are a recurring motif that is often used as a metaphor for gaining knowledge about oneself and God. This image is an example of this architectural element found in a byzantine church. Photo taken by the Author

In the name of the rose, the labyrinth represents a world containing all knowledge only accessible to those who have been initiated. In the Middle Ages, the concept of the labyrinth was central and usually represented a difficult and confusing path *associated with the deepest knowledge of God and the universe* [2].[1]

Like William of Baskerville, we learn to navigate natural and complex systems as if they are labyrinths. In fact, science has developed methods to find the best way to navigate each maze. In this way, it is possible to map any complex system into a clear structure by breaking it down into simple and clearly observable and coherent components. This requires the implicit assumption that systems are complete, with clear causal paths, and therefore mathematically representable.

Hence, the concept of completeness implies a fundamental common structure that rules nature's complex systems. This is equivalent to assuming a single architectural principle for any labyrinth. Thus, using mathematics and causality concepts, it is possible to find the right path in the middle of several corridors and cul-de-sacs.

With this in mind, science is like Ariadne's ball of thread (ο Μίτος της Αριάδνης, "Ariadne's string"), in which she gave to these balls to retrace his way out of the Minotaur labyrinth: like the ball of thread, science provides us with tools to understand the fundamental principles and architecture of the labyrinths to help navigate us out of them.

In complex systems, the labyrinth is made up of several interconnected elements that group together and build multiple scales. These are complicated physical systems, such as liquids, gases or even polymer melts. This method is also applicable to biological and social structures, such as sophisticated cell mechanisms or flocks of

---

[1] This deep knowledge was called gnosis.

birds. Penetrating all these labyrinths requires patience and perseverance to discover the common patterns that lead to their shared underlying architecture.

In recent decades, an increasing number of mazes have been solved because of significant advances in mathematics. Consider the mathematical description of quantum mechanics and its pragmatic approach in describing the quantum world [1] or Mandelbrot's work on chaotic systems [4]. These concepts are helpful in understanding physical phenomena such as phase transitions, as well as characterizing complex dynamical systems and their self-organization [7].

This awareness is useful in understanding other systems outside classical physical systems. In such systems, extreme events are likely to be observed with drastic jumps, and their distributions cannot be represented with a simple normal distribution (which is usually associated with regular events). Instead, there is chaotic behavior, where similar fluctuation patterns can also be observed at different scales (scale invariance), which is the characteristic behavior often observed in biological and physical systems.

This led to the introduction of a scale-free distribution, i.e., the larger you increase this structure, the more likely you will come across similar patterns observed in the beginning. A rather groundbreaking concept associated with fractals and chaos theory is often applied to risk analysis in finance [4].

This meant that the discovery of scale-free distributions is similar to the architectonic principle, thus helping us understand the structure of complex chaotic systems by giving us the option of predicting the system's behavior.

Therefore, we attempt to collect accurate facts from nature and the universe via empirical methods and by dissecting complex systems more precisely. Such methods, combined with theoretical concepts about the fundamental principles governing any complex system, some of which are based on the concepts of causality, allow us to develop tools such as chaos theory in developing predictive models.

However, some complex systems are still unpredictable. In such complex systems across multiple scales, incompleteness, i.e., difficulty in deriving a coherent system description, persists, making causal inference difficult (or, in other words, our understanding of complex systems is necessarily incomplete [6]). For example, mathematicians are still struggling to obtain an accurate stock market model. In addition, crises and crashes are becoming increasingly difficult to predict. The same is applicable here, as biological systems still remain a challenge, which explains why it is extremely difficult to recommend personalized therapies to cancer patients as an example, despite all the advances in biomedicine.

Therefore, is it true that the principles of complex systems follow a single architectural principle? Or are there natural limits restricting the search for consistent causal pathways in complex systems, thus limiting our discovery of deterministic laws? In this case, synthetic systems do have fundamental limits.

As a result, we are aware of the fact that the world appears incomplete, as if there were a labyrinth within a labyrinth, and that solving a labyrinth creates new unforeseen labyrinths within it.

This book introduces techniques and methods to recognize a "labyrinth within a labyrinth" and how we can analyze the limits on a system's observability from the following three perspectives:

1. The first is a structured analysis and overview of theories of causality in complex systems on the basis of physical principles. In addition, we investigate how the interpretation of causal events and causal properties leads to the definition of rules, principles, and laws in well-defined systems. This approach is therefore useful for those interested in systems theory from philosophical and theoretical perspectives. However, this topic is widely broad. To this end, in this review, we focus on the most important concepts related to causality and determinism in systems theory.

2. The second perspective is causal inference. The aim is to introduce novel techniques based on topological persistence to perform causal inference. We use this method for feature engineering in statistical modeling. This topic has a practical nature and should be useful for performing statistical modeling of feature extraction in time series.

3. The third perspective is the introduction of a theory of elastic states and persistent incompleteness in complex systems. The goal is to recognize how complex systems can actually have more complex dynamics and methods, such as modeling complex systems via complex networks, or concepts of emergence that need to be revised. We also examine fundamental aspects of cognition, arguing that cognition is not just a byproduct of emergent systems. Multiscale systems are likely to require basal cognition, which is related to basic decision-making.

While the first two perspectives are pragmatic and could be useful in providing practical tools to those who implement mathematical models, the last perspective aims to introduce novel and speculative theoretical approaches. It offers a fresh approach in looking at the foundations of complex systems and seeking answers to difficult questions such as the nature of consciousness and life.

Furthermore, this last perspective implies that prediction cannot be reduced to a single cognitive space relative to well-defined contexts (as is the case with modern fundamental models such as large language models [5]) and that statistical outliers are extremely relevant to understanding how incomplete the system is.

These three perspectives are summarized in six chapters, with a focus on systems theory applied to complex systems, taking into account the elastic states.

In Chap. 2, we provide an overview of the philosophy of causality and causal determinism in systems theory.

In Chap. 3, we embark on a more technical overview of mathematical modeling in systems theory. We focus on deductive modeling (from the perspective of the natural sciences), inductive modeling (from the perspective of machine learning), and causal inference.

In Chap. 4, we present the mathematical theory required to understand elastic states in the context of incompleteness in multiscale systems and its implications for modeling. In Chap. 5, we proffer the influence of elastic states on causal inference and introduce appropriate complexity measures to better understand such cases. In conclusion, in Chap. 6, we discuss the third perspective and examine the effects of elastic states on the understanding of complex systems.

This last chapter is not just a summary, but an exploration of how basal decision-making is relevant to the definition of complex multiscale systems. All these chapters are complete on their own, so it is not necessary to read this book from the beginning to the end; rather, the reader could decide which chapter to delve into according to his/her topic of relevance and interest.

The greatest challenge of this book is its multidisciplinary approach, as it is written for readers with backgrounds in physics, systems theory and systems biology, biology, sociology, economics, informatics and philosophy. Topics and terminology can have different meanings depending on the reader's perspective, educational background, cultural exposure and willingness to remain open. Thus, it can be very demanding.

For example, in Chap. 2, we introduce the concept of the "small world", which is interpreted as a reductionist approach and has no direct reference to the concept of small-world networks.

Additionally, the concept of networks is interpreted differently depending on the reader's background; here, a complex network (commonly used in physics) is equivalent to the mathematical concept of a graph.

We also consider complex systems with interacting elements; here, an element is a general definition that can refer to both a particle or an active agent (like organisms or individuals). We provide the corresponding definition depending on the theoretical interpretation of the system in each chapter.

In general, we refer to complex systems from a universal perspective. However, due to the nature of the topics in this book, there is a fluid relationship among general complex systems (consisting of interacting physical elements), biological systems (consisting of molecules that regulate cell signaling or cell metabolism, or generally composed of organisms) and/or social systems (consisting of agents).

As we argue in Chap. 6, we contend that basal cognitive properties are shared by multiple complex systems, especially when complex systems are labyrinths within labyrinths.

In this book, we are not trying to present a theory to make the world observable and controllable by solving any labyrinth. Rather, we are purporting to recognize the existence of a labyrinth within a labyrinth. We have also realized that the search for meaning in the world cannot be ignored (a search that may also be shared with elements within systems), which raises possible limitations to a utilitarian approach in understanding nature.

References

1. Baggott J (2024) 'Shut up and calculate': how Einstein lost the battle to explain quantum reality. Nature 629(8010):29–32. https://doi.org/10.1038/d41586-024-01216-z

2. Doob PR (1990) Etymologies and verbal implications. In: The idea of the labyrinth from classical antiquity through the Middle Ages. Cornell University Press, pp 95–100. https://doi.org/10.7591/j.ctvn1t9v6.10

3. Eco U (1982) Der name der rose, 1. Aufl. Translated by Kroeber B. Deutscher Taschenbuchverlag, München

4. Mandelbrot BB (1997) Fractals and scaling in finance. Springer, New York, NY. https://doi.org/10.1007/978-1-4757-2763-0

5. Millière R, Buckner C (2024) A philosophical introduction to language models—part I: continuity with classic debates. arXiv https://doi.org/10.48550/arXiv.2401.03910

6. Siegenfeld AF, Bar-Yam Y (2020) An introduction to complex systems science and its applications. Complexity 2020(1):6105872. https://doi.org/10.1155/2020/6105872

7. Suteanu C (2022) Scale, patterns, and fractals. In: Suteanu C (ed) Scale: understanding the environment. Springer International Publishing, Cham, pp 207–252. https://doi.org/10.1007/978-3-031-15733-2_7

Chapter 2
From Small Worlds to Systems Theories

Keywords Causality · Causal determinism · Philosophy · Small world · Large world

This chapter begins with a discussion of systems theory in relation to physical and biological systems and their philosophical foundations. Before embarking on more technical questions, what makes philosophy so important?

Systems theory has become an important tool for gaining a comprehensive and fundamental understanding of complex systems throughout the universe, ranging from large structures (such as galaxies) to biological systems (such as viruses) to those of the nervous and brain systems.

Thus, any theory that spans many systems and addresses our self-existence has philosophical roots.

From a philosophical point of view, systems theory is a form of structuralism [5] that is oriented toward the concept of the "small world". According to this concept, systems theories are derived from simple and well-characterized interrelated and fundamental events that are systematically upscaled and provide solutions to difficult problems in various fields, such as physics, biology, and even social and cognitive sciences.

The extraction of "small worlds" in this context is part of a systematic and methodological approach: dissecting complex systems and isolating interacting elements, extracting their fundamental properties and causal interactions, and systematically reconstructing large systems from these simple interacting elements.

However, this approach can be extremely restrictive and excessively reduces the understanding of the fundamental states of the system. In this chapter, we provide a conceptual overview of the benefits and challenges that the small-world approach brings to systems theory in general.

© The Author(s), under exclusive license to Springer Nature Switzerland AG 2025
J. G. Diaz Ochoa, *Complexity Measurements and Causation for Dynamic Complex Systems*,
Understanding Complex Systems, https://doi.org/10.1007/978-3-031-84709-7_2

We also discuss the theoretical background of deterministic causality and its relationship to the concept of the small world and its role in systems theory.

2.1 What Is Causality? A Brief Overview[1]

Depending on the discipline, there are specific interpretations of what a *systems theory* should be. In this book, we address *general systems theory*. This approach has strongly influenced not only the understanding of biological systems but also the development of control theory [1, 9, 52]. A central concept in this theory is causality. In this section, we provide an overview of causality, not only from a purely technical perspective but also from a philosophical perspective.

2.1.1 *Correlation, Causality and Human Culture*

Humans are characterized by their ability to observe relationships between events that are essential to developing fundamental knowledge about nature and the environment. Correlation simply means establishing a relationship between two observations A and B. However, we tend to extract causal notions from these observations. This causality means that if I can identify an element associated with an event A that eventually triggers an event B and if this observation can be repeated numerous times, then there is an underlying interaction and information transfer between Events A and B. As a result of this knowledge, we can see that Events A and B are intertwined by some underlying natural principle.

Consequently, the ability to control Event A implies that Event B could also be controlled. For example, how much Novocain a patient receives depends on how much pain they feel during dental surgery. A causal concept may also be referred to as metaphysical when two correlated events are connected by unseen mechanisms.

According to the modern scientific paradigm, the mechanism by which an event triggers another constitutes the basis of physical and natural laws. Knowledge of causal relationships is one way to understand how a system behaves deterministically.

The concept of causality is an essential part of the human soul. Our ancestors had a very good perception of causality (and not just mere observation of correlations) that led to their survival for hundreds of thousands of years in different ecosystems and environments, eventually leading to the discovery of the agriculture, engineering, and principles necessary to build complex societies.

---

[1] This one is a fast overview about causality. For a more deeper discussion about causality and its technical application in different fields look into this overview about causality: Statistics and Causality: Methods for Applied Empirical Research—https://onlinelibrary.wiley.com/doi/book/10.1002/9781118947074.

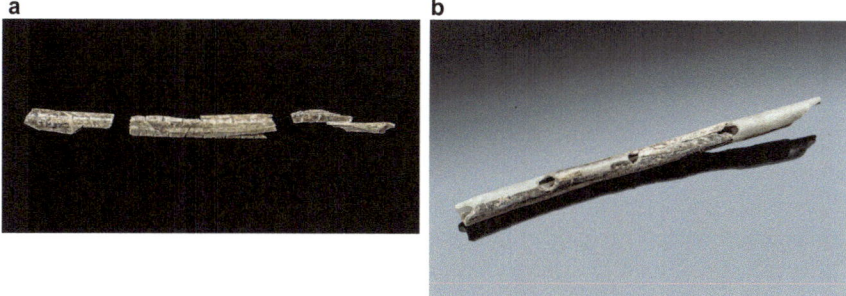

Fig. 2.1 Human culture is characterized by deep knowledge of the causal relationships among different phenomena. A sophisticated knowledge of physical relationships is helpful in developing complex instruments such as flutes. This is one of the ancient examples: a 40,000-year-old flute found in the Klösterle cavern near Blau Beuren in Baden Württemberg, Germany. **a, b** The different restoration phases of this flute are shown. Archäologisches Landesmuseum Baden-Württemberg/ Foto: Manuela Schreiner

An intuitive knowledge of causal relationships has helped our ancestors develop sophisticated technologies, such as musical instruments (e.g., Neolithic flutes, such as the one shown in Fig. 2.1, which is approximately 40,000 years old and is the oldest musical instrument ever recovered in an archaeological excavation), complex Neolithic stone structures that are more than 6000 years old (which require thorough knowledge of physical principles for their construction) [55], or the domestication of plants and animals [27].

Causal relationships are not limited to the physical world. Humans also have the ability to recognize metaphysical relationships between different events. Clearly, such relationships between the cosmos and humanity cannot be interpreted as real causal relationships. They have to be seen simply as correlations.

Examples of such metaphysical causality include the relationship between natural phenomena and fate—the position of the stars and their relationship to the life cycle of the environment were observed with high precision in the Sumerian cultures to keep a record of periodic events that correlated with the position of planets and stars to periodic events on earth[2]—in what can be perhaps the first empirical and data-based approach to understanding patterns in the environment and the world [36].

This aspect is so important to human character and culture that civilizations have put great effort into constructing temples, myths, and religious systems that aim to estimate, predict, and even correct such metaphysical relationships through rituals, from prehistory to modern times [27, 64].

Certainly, there is invaluable knowledge, deep wisdom and understanding of nature in such mythologies, which have not been fully evaluated and recovered; rather, they have been consequently ignored, resulting in irreparable loss in the

---

[2] https://www.cambridge.org/core/books/cambridge-history-of-science/science-and-ancient-mes opotamia/C48D6E70188ED938863F479B692D465B.

process similar to a kind of cultural evolution and natural–cultural selection, for example, due to constant human migration [34].

Intuitive notions of causality or relationships do not necessarily imply a formal definition of causality. This was embraced only by the founding of philosophical schools in Greece. Even in the early days of modern philosophy, the mythical interpretation of causality had a less direct influence on the formal construction of logical systems and philosophy; nevertheless, the indirect contribution of myths to this formalization, which has guided the development of modern culture over the past 3000 years, should not be ignored (see, e.g., *from Myth to Reason? Studies in the Development of Greek Thought* [14, 16, 45]).

2.1.2   First Philosophical Foundations of Causality

Causality is not only an integral part of our culture but also a fundamental part of our idea of understanding nature and our existence in the universe and the universe as a whole. Therefore, causality is at the heart of philosophy, as it is a way of deriving logical relationships from nature and the physical world to expand our knowledge of this world [65].

Aristotle is perhaps the most influential Greek philosopher on the subject of causality. Aristotelian science studies a particular aspect of reality, and in this way, causal knowledge is obtained, i.e., the knowledge of relevant or appropriate causes. As a result, Aristotle developed a theory of causality known as the Doctrine of the Four Causes [23].

I. *The material cause* or that which is given in reply to the question "What is it made out of?" What is singled out in the answer need not be material objects such as bricks, stones, or planks. From Aristotle's perspective, A and B are the material causes of the syllable BA.

II. *The formal cause* or that which is given in reply to the question "What is it?" What is singled out in the answer is the essence or what-it-is-to-be something.

III. *The efficient cause* or that which is given in reply to the question: "Where does change (or motion) come from?" What is singled out in the answer is the whence of change (or motion).

IV. *The final cause* is "What is its good?" What is singled out as the answer is the purpose in which something is done or takes place.

For Aristotle, a firm grasp of what a cause is and how many kinds of causes there are, is essential for a successful comprehension of the world around us [23]. It was crucial in the development of mathematics to have such an original understanding of the cause. For example, in Euclid's Elements, one of the most influential mathematical texts; it has long been a tradition to read Aristotle's treatment of the first principles first to better understand basic mathematical postulates.

However, this notion of causality proposed by Aristotle hindered scientific development and set standards that were considered dogmatic rather than fact, such as the interpretation of eternal harmony represented by circles (which led to the development of epicycles to explain the orbit of the planets).

Therefore, a certain amount of caution is required when considering Aristotle's concepts of causality today.

Modern philosophical and scientific thinking has been inspired by such perspectives. Consider, for example, Descartes' claim that the cause of the object must contain many realities, such as the object itself, or David Hume's empirical definition of causality, which aims to answer two fundamental questions: How do human beings obtain the idea of causality? How do human beings infer effects from causes and causes from effects? [6, 53].

Such developments have drawn attention to the problem of causality to systematic experience from the physical world, i.e., to the Aristotelian material cause (1st form of cause), as a way of deriving an objective and absolute concept of cause throughout the universe.

Perhaps a relevant exception in this tradition is provided by Kant, who categorizes causality as a mental construct, implying that causality is not only objective but also depends on the observer[3] [15]. This approach contrasts with the purely objective idea of causality and implicitly involves a type of subjective perception as well as a mental state that is included in the notion of causality (a concept developed by E. Mach [20]). This is contrary to the conventional view, which claims that every process can be characterized objectively and empirically.

2.1.3 Logical Positivism and Physics in Causality

Logical positivism places systematic empirical observations at the center of its philosophy [41]. In this philosophical approach, the verification principle is fundamental, denying the existence of an a priori principle on the basis of an empiricist criterion [60].

This philosophical approach has guided the development of a contemporary scientific perspective that includes physical theories to discover such causal relationships. Neither passive interpretation nor gentle negotiation of one's own future are the goal. The main goal of deriving causal relationships that can be empirically verified is to empower individuals to actively control their world. This helps them design and construct their future. To this end, causality is often paired with the predictability and deterministic behavior of the system, two aspects that are often confused or mismatched [4].

The importance of causality in light of the theory of relativity and quantum mechanics is also relevant here. Whenever a signal is exchanged between two events, the theory of relativity claims that the effect does not precede the cause. Since nothing can exceed the speed of light, this limit sets a natural limit to causality. In particular, a result cannot occur from a cause that does not lie in the rear (past) light cone of

---

[3] Such notion will be relevant in Chap. 6 in this book.

this event. In the same way, a cause cannot have an effect outside of its front (future) light cone [47].

On the other hand, however, the development of quantum extensions of causal models has proven difficult owing to the special features of quantum mechanics. For example, if two or more quantum systems are entangled, it is difficult to deduce whether the statistical correlations between them imply a cause–effect relationship [2, 46]. Again, the concept of causality is restored when quantum processes are considered unitary, i.e., when they preserve quantum information.

2.1.4 Impact of Concepts of Causality

The notion of causality and determinism is currently intrinsic in part of our culture and constitutes an essential part of how to interpret reality and develop technologies. Using concepts derived from the notion of causality, we can construct the theoretical basis of fundamental mechanisms by systematically observing natural systems and extracting from these observations different events that are not only correlated but also interdependent and exchanging information [7].

With such observations, the ultimate aim is to determine under which conditioned systems behave deterministically and could then be controlled.

However, this concept of causality is not universal. Several other cultures that have different traditions than the European tradition have also developed other concepts of causality. These include Asian or ancient American cultures, which advocate a relativistic and organic conception of causality, in which the entire context influences interrelated events.[4]

Nevertheless, the prevailing concept of causality has been predominantly influenced by European culture and, more recently, by logical positivism. A central aspect of logical positivism is the concept of causal determinism, which is essential to systems theory. In the next chapters, we will take a closer look at this aspect.

2.2 Causal Determinism and Observables

Logical positivism emphasizes the need to collect enough data from observed elements, as well as their interrelationships in space and time, in a consistent way to avoid any metaphysical arguments [37]. In this way, evidence of their existence in the present and in the future is collected.

Assuming that the system is deterministic and that there are causal interactions between two entities, it is always possible to claim that when A changes its actual state and interacts with B, B also changes its state. Therefore, we can prove only the

---

[4] See for instance this short blog article: https://science.jrank.org/pages/8561/Causation-in-East-Asian-Southeast-Asian-Philosophy-Influence-Confucianism-in-East-Asia.html.

existence and connection of events through precise empirical observations [21]. In general, and strictly speaking, theories cannot be verified; they can be confirmed or refuted only to a certain confidence level [41].

Complex systems need a clear framework to fulfill this logic-positivism. To this end, it is necessary to introduce the concept of a statistical distribution. Complex systems usually involve sampling several observed elements at different scales. This method is preferable to the detailed description of the state of each element, which is computationally impossible. Therefore, statistical distributions can be useful in describing systems with many elements. Thus, understanding causal determinism requires an investigation of how it relates to statistics.

In the next section, we introduce the notion of causal determinism and provide an overview of its relation to statistics, particularly statistical mechanics.

2.2.1 Causal Determinism: Initial Remarks

Any attempt to provide an accurate definition of causality requires a basic notion of completeness. This helps to understand the underlying structure of a system perfectly. That is because the notion of causality is linked to the idea of determinism [30]. Thus, knowing exactly how an event leads to the next event, or a series of events, it is possible to predict future scenarios. And this represents Power.

Some literary works demonstrate this form of power. In Mark Twain's novel "The Yankee at Arthur's Court", the hero Hank Morgan accidentally travels back in time and ends up at Arthur's court in the Middle Ages. Because of the court's magician, Merlin, and his intrigues, the hero, Hank, is imprisoned and given a death sentenced. However, he is smart and knows about solar eclipses. Therefore, he pleaded the king to hold the execution to the day and hour where an eclipse would occur.

The hero saves his own life, demonstrating that he is able to control the sun when he orders its sudden disappearance on his execution day. By doing so, Hank embodies being the most powerful magician in the world. The hero's clever plan exonerates him and proves his power[5] [59]. In this context, the deterministic causation of a solar eclipse, i.e., the transit of the moon in front of the sun, casts a shadow on the Earth's surface, soon became a type of power for the Yankee.[6]

Our desire to discover causal determinism in nature is rooted in the belief that any system, including living systems, can be objectively observed and identified by understanding its causal relationships in an objective manner. And when condensed in the following relationship, we call it the OACEM relationship:

---

[5] The same story has been copied by Hergé's Tin Tin in the Inca empire. In this story, the hero saves himself by correctly predicting a solar eclipse on the same day as his ritual sacrifice.

[6] https://etc.usf.edu/lit2go/174/a-connecticut-yankee-in-king-arthurs-court/3049/chapter-6-the-ecl ipse/.

(Observe the system) {O} → *(Analyze the **C**ausal pathways between the elements/*   (D-1)
events in this system) {AC} → *[Extrapolate different systems] {E} Λ [**M**odify*
them systems] {M}.

The relation (D-1), which is the essential foundation of physics, chemistry and every discipline rooted in systemic concepts, has its roots in the fundamental theories of Democritus and Aristoteles[7] [23]. The aim of this relationship is to create the basis for an ontology, i.e., the conceptual and philosophical basis of a system that takes care of the description of an objectively describable world (without relying on a subject or an actor) and the assessment of the existence of objects and concepts [32].

The best way to develop a successful natural science theory from a philosophical perspective is to avoid asking the question "why" (rooted in metaphysics) but rather to ask only the question "how" (rooted in epistemology). This statement is often attributed to Einstein, but it is based on epistemology and the avoidance of metaphysical arguments in natural science [38, 57].

Nevertheless, not every event in the world follows the same trajectory as those defined in classical mechanics, such as the trajectory of a planet or satellite in celestial mechanics [26]. Determination and knowledge about the future are often difficult, if not impossible. The simple fact that Aristotle presents a doctrine of four forms of causality suggests that there is no single coherent definition of causality. Depending on how a system is analyzed, different forms of causality arise. While all these interpretations seem disjointed, they are actually deeply connected.

The problem of causality is relevant in the age of big data: the question arises of how causal relationships can be inferred from observational data. Thus, from the classical extraction of primordial laws governing fundamental interactions, the availability of large amounts of data collected across multiple scales has opened the door for the systematic integration of events that are theoretically related to stable correlations with underlying causal pathways that can be statistically analyzed and placed in the context of a formal causal theory [15]. This concept is the basis for network theories applied to complex systems [12].

Therefore, when determinism is considered, we are not simply dealing with individual punctual events that can be causally linked to each other, such as collisions between two particles. Instead, we address statistical and ultimately chaotic systems with individual microscopic nondeterministic events. Despite the apparent impossibility of deriving causal paths from individual events, the main goal is to define statistical distributions that can be represented as events and to describe these distributions in terms of causal determinism.

This aspect is essential from two perspectives: first, the relationship between statistics and deterministic causality, and second, the use of statistics to extract fundamental relationships to describe complex systems.

Regarding the first perspective, the goal is to derive knowledge from empirical data when, under special conditions, probabilistic distributions are connected to causal

---

[7] https://plato.stanford.edu/entries/aristotle-mathematics/supplement1.html.

networks [43]. It represents a special chapter in the theory of causality, since it uses empiricism as a fundamental way to derive causal relationships, according to logic-positivism [60].

The second perspective is fundamental for understanding complex systems in which not only the connection between events but also the relationship across different scales is of central importance too. To do this, it is necessary to consider the statistics of complex systems.

2.2.2 Causal Determinism, Statistics and Observables

As we introduced in the first part of this chapter, the simple idea of causal determinism is to acknowledge that a single event A triggers a next event B such that $A \rightarrow B$. If this regularity can be demonstrated in several observations, then every time it is observed, it should be possible to observe what takes place. Setting such events as simple observables in space and time is the simplest and intuitive form of causality [15].

In reality, however, we are dealing with systems that are made up of many variables that can only be described with statistical distributions. Therefore, an event A cannot simply be defined as a point in space and time because there is an available collection of events that need to be sampled either in space or in time or both.

In this case, it is not relevant to track individual elements in the system but rather to describe the state of the system Ψ, which is defined by the average of all the states of the individual elements. Any change in this state, e.g., owing to the change in a physical parameter or the fact that inherent symmetries determine the form of that function (e.g., by minimizing a control function), is thus the representation of the interrelation of one event to another event in causality.

In this way, the evolution of the distribution function Ψ can be estimated.

Owing to their deterministic character, the formalization of distribution functions provides the basis for the formalization of causal determinism in the physical sciences and complex systems [58].

The relevance of this concept is restricted not only to classical systems but also to quantum mechanics. According to the Copenhagen interpretation, understanding all the underlying processes behind quantum phenomena is not relevant. Instead, the main goal is to represent a function Ψ representing all the possible states of the system [40].

Perhaps the most relevant point is to avoid any classical interpretation, such as representing or abstracting elements in quantum systems as classical particles, and to understand quantum states as probability distributions[8]; by doing so, it is possible to consider the effect of the observer on the system.

---

[8] "A common misunderstanding in quantum mechanics": This article is relevant to better understand the a common factor of discussion and confusion in quantum mechanics https://www.chemistry world.com/opinion/a-common-misunderstanding-about-wave-particle-duality/4019585.article? utm_source=Live+Audience&utm_campaign=9769b6ec30-nature-briefing-daily-20240619& utm_medium=email&utm_term=0_b27a691814-9769b6ec30-49911868.

Hume assumes that it is possible to observe events objectively through systematic observations and that any access to the reality of the universe can be recorded in a more or less objective way.

However, in quantum mechanics, any subjective involvement must be taken into account. For the first time, it is possible for the observer to have a certain influence on the observed system. Therefore, the reality of the universe does not seem to inherently exist by nature and depends very much on different observers.

Despite advances in quantum computing, the problem of quantum mechanics interpretation is a much-discussed topic that still affects the way quantum systems can be constructed, with theories such as decoherence theory providing underlying deterministic interpretations using pilot waves, helping to derive an underlying deterministic nature that is distorted when classical scales interact with the quantum world [49, 56].

Therefore, all these problems indicate that statistics and statistical interpretations of complex systems do not exist inherently and that there is some kind of epistemic component in their construction [65]. This is a deep contrast to the program proposed by logical positivism, since it excludes the possibility of extracting universal logical relationships. In addition, depending on the system, the observer should be considered (e.g., with the help of Bayesian statistics and epistemic causality [65]).

The different scales are also relevant. Considering, for example, the way cancer is understood, there is currently much focus on how genetics might be responsible for the expression of cancer (and how genes determine the synthesis of proteins that lead to a mutation of healthy ones into cancer cells). Other influencing factors, such as the environment and nutrition, are often ignored because of this focus. Now, all these factors are ordered on different scales; however, their integration into a causal model is far from trivial. In such a case, creating a deterministic causal model by integrating all these factors into causal paths may lead to a mathematical model that is extremely difficult.

All these aspects point to the problem of the isolation of factors and observables in deriving deterministic relationships. In addition, the observation of single events and the derivation of apparent causal pathways can be misleading, implying that empiricism should not be the final test used to validate a theory [15].

Although some scholars are critical of concepts such as causality and determinism, the way systems theory is applied and used for practical purposes takes a completely different path. This is motivated by the perspective of synthesizing biological processes and living organisms. This goal, inspired by the philosophy of Hume, has tremendous consequences, not only in terms of how science is done but also in terms of how science is translated into technology [48].

2.3 The *Small-World* Concept and Systems Theory

Statistical distributions objectively represent the state of the system, regardless of the position in space and time of the observers measuring it. The observer can disturb the system and bring about a change in the state ψ; however, the simple and passive act of observation should not produce any change in the observed system or its state Ψ.

To define such functions, it is necessary to isolate them in the best way. A small-world concept is a method of observing an observable (which should not be confused with the concept of small-world networks). Essentially, this reduction is similar to the dissection of natural systems: science tries to find the simplest elements that constitute its nature so that it can be controlled.

According to this concept, the aim is to define observables in "small worlds", i.e., observables that are located in an environment without surprises, which condenses all the fundamental principles of the system. These principles are similar to the basic axioms (or basic building blocks) required to characterize the system.

In real "large-world" problems, probabilities are difficult to define, and observables are subjected to unexpected and contingent fluctuations due to their interactions with the environment (high uncertainty); "small-world" systems, on the other hand, are perfect and ideal cases with low uncertainty, where observables, probabilities, and parameters are perfectly defined and controlled [62].

For example, instead of describing the physics of polymers by considering all the properties of single molecules or even atoms, it is much more practical to isolate monomers, i.e., they are considered perfect suspensions without interactions with other solvents or the environment and generate coarse grains by representing them as spheres (see Fig. 2.2).

Thus, monomers constituting a polymer resemble pearls in a collard [11]. This reduces any, if not all, potential complexity in the system and preserves all key aspects of the system. By reducing the system to its essential and simplest elements, the most basic information about the system can be provided.

The definition of such a thing would be extremely practical, as it would require perfectly controlled conditions that can be reproduced experimentally. Moreover, such a reduction is accessible from a theoretical point of view, since it allows the definition of a system that eventually follows clear rules that can be related to fundamental principles that would be universal in principle, i.e., would be valid in every corner of the universe.

The decomposition of a complex system into small worlds is therefore essentially the way to distill the fundamental principles of the universe from complex phenomena.

Descriptions of small worlds also suggest that "big world" problems can be derived from "small world" descriptions. According to this theory, descriptions of "small worlds" are connected in various ways. "Large worlds" are therefore the result of this networking and its dynamics. As a result of this concept, large worlds are causally determined by small worlds.

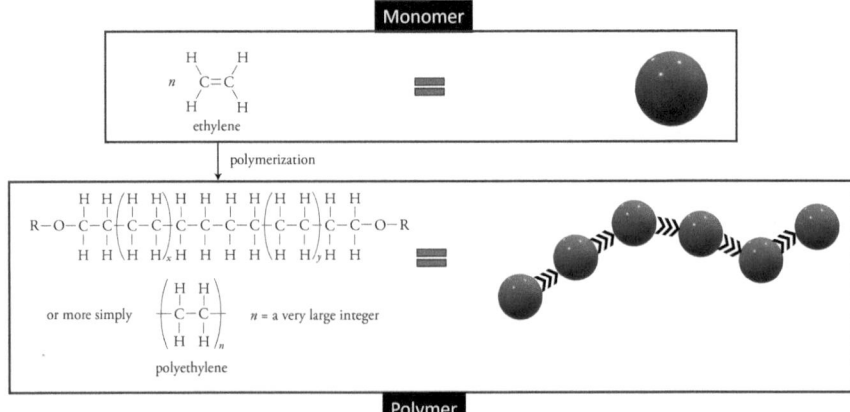

Fig. 2.2 Using coarse-graining of the physical–chemical structure of monomers as spheres and joining them with collars as an example of the small-world concept. Thus, polymer dynamics can be developed as the sampling of these collars while considering different physical constraints (boundary conditions and external reservoirs) as well as physical parameters (such as temperature)

A limitation of such an approach is the presence of aleatoric noise and epistemic uncertainty. This refers to the inaccuracy of observed events caused by effects such as measurement inaccuracies or a lack of visibility of the measured data.

However, it is possible to take this into account as part of the modeling process (for example, using Bayesian statistics; for a complete review of the consideration of aleatoric and epistemic uncertainty, see, e.g., Hüllermeier and Waegeman [33]).

This way of thinking is very practical from a systemic point of view, as it allows for the systematic upscaling of rather simple elements. This makes it possible to systematically increase the complexity and sophistication of the system via simple physical concepts.

The systematic exploration from less complex to more complex systems can be traced back to the laws of physics: the Cartesian mechanistic universe was only possible after the discovery of simple laws of mechanics by Galileo and Newton [39]. It has produced physics-relevant results, ranging from understanding quantum mechanics to understanding complex systems such as polymer melts or weather systems.

For this reason, scientists start by studying very simple systems or toy models before trying to understand the complexity of a system as a whole (see Fig. 2.3). The impact of this methodology can also be felt in the way engineering is performed and how new paradigms are emerging from the engineering and development of new devices, which has guided the development of new scientific knowledge [29, 39, 61].

For example, a technical system such as a computer can be reverse engineered, and from systematic analysis to switching on and off various elements, the correct connections and the final function of the device can be derived. From this, not only is the function of the simple interacting elements but also the function of the device

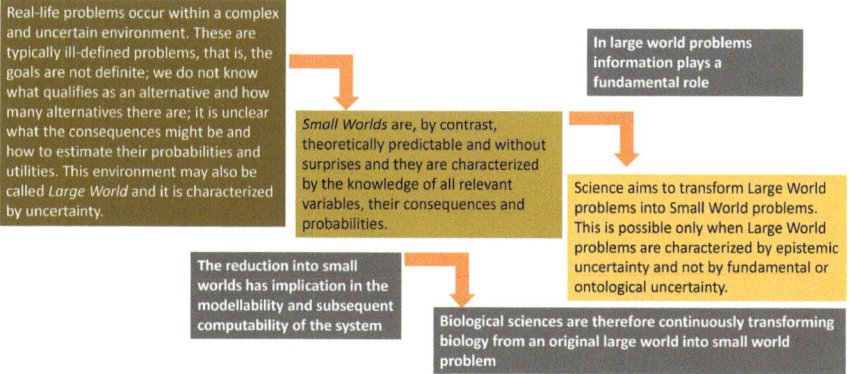

Fig. 2.3 Logical schema for the reduction of complex systems into small-world systems, following the concepts of large [50] and small worlds (as postulated by Viale [62])

itself can be derived. Therefore, reverse engineering is relevant in both engineering and science, since by applying a similar methodology, it is possible to extract the basic elements of complex technical devices, as well as living organisms and cells [63].

By reversing the engineering of gingival pathways and perturbing gingival pathways, it may be possible to determine the function of the cell and its key proteins involved in signal transduction or in metabolomics [63]. As a result of these concepts, which have become the standard in systems biology theory, they are currently guiding how the way scientists seek to gain a deeper understanding of living systems, their function, and their relationship with other organisms, with the aim of being able to "hack" them (which has sparked a scientific and cultural trend in which citizens outside academia aim to DIY biology (do it yourself) and optimize themselves and thus modify their own organisms [44]). This means modifying fundamental pathways to cure diseases, improve crops in agriculture, or even optimize biological functions to discover the elixir of eternal youth.

2.3.1 Small World, Completeness and Causality

The mathematical advantage in reducing complex systems into "small worlds" is to have a system that can be controlled, with states that can be reproduced in any new experimental setup. This guarantees that the system can be observable and perfectly characterized, which is helpful for deductive modeling and mathematical characterization.

In addition, it can be assumed that the modular combination of simple systems can lead to both the construction and further development of more complex systems. Such simple principles, which can be used in the definition of basic and complete mechanistic models, are therefore fundamental axioms for the derivation of any complex system. This concept can be summarized under the following definition:

In respect to a particular property, every mathematical formulation of a complex   (D-2)
multiscale system can be derived using single small-world systems.

This definition, which is for complex systems, is similar to Hilbert's idea to derive a formal system [67]; it can essentially be seen as a formalism for the derivation of complex systems: axioms (which are the result of the decomposition of complex mathematical concepts) are the basic building blocks in deriving any mathematical concept.

It should also be possible to build any complex system from fundamentally decomposed small worlds. For this reason, a fairly simple system can be used to explain and understand several complex systems from physics to sociology [10, 24].

Following definition (D-2), the observation of a system S implies observing its state Ψ_S. If such a system is isolated, if the observed states observed in different experiments, i.e., $\{\Psi_S^1, \Psi_S^2, \ldots \Psi_S^i\}$, are almost stable and can be reproduced in new observations or experiments. Furthermore, if the system S is identified within *a small-world* context and can be considered a rather simple module for more complex systems, then it is possible to affirm that this system is a basic axiom that can be used to derive the mathematical definition of the whole system.

If we ignore the logical difficulties that the small-world concept entails (i.e., the extreme simplification of inherent structures as well as the interaction of the system with its environment), it represents the perfect way to establish causal relationships in complex systems: either the change in a system's parameter triggers a change in the state Ψ_S, or one state Ψ_S^1 can trigger a change in a second state Ψ_S^2. Thus, causality is intrinsically related to the *small-world* concept.

Thus, the identification of *small worlds* fulfills several relevant porpoises in systems theory (from both theoretical and experimental perspectives):

- It provides a basic formalism that guides the initial definition of a system. On this basis, it is possible to define modular mathematical models used in computer simulations that can eventually be qualitatively or quantitatively validated. Modeling based on exact formal mathematical definitions is a deductive approach.
- The definition of small worlds is helpful in describing the fundamental properties of a system and can be used as a method for experimental design. Thus, such a method is not used to obtain information but rather as a guide to define the relevant observed states Ψ_S^i that define the system, i.e., it is a method to define the data necessary to characterize the observed system.
- By assuming inherent and interacting simple elements, characterized as small worlds, it is then possible to perform deductive modeling. From the obtained data, we attempt to deduce a mathematical characterization of the model.

However, complex systems are not purely mechanistic: they are characterized by both physical responses and functions, i.e., the role they play with other systems in a wider context. Such a function is thus a specific output in relation to the system's environment or other systems.

Therefore, there is a paradox in the formal way in which complexity is first reduced in a small world before it is systematically derived from a large world. In relation to the Aristotelian classification, consider the interrelation between matter (1st category) and the final cause (4th category): cell regulation can, in principle, be explained as the interaction of several chemical substances that can be represented by complex networks.

However, this material is fundamental because chemical interactions within the cell maintain its function and regulate its interaction with the environment. The material cause thus determines the root cause and vice versa. This aspect represents an important limit in our pursuit of more knowledge, which we gain with causal relationships.

The success of the *small-world* concept in causal determinism is the reason why it continues guiding the mathematical description of such systems. Nevertheless, there are practical limitations that compromise such theories.

2.3.2   Small Word Concepts and Systems Theory

According to the previous definitions, there is a constant combination of data and theoretical formalism that helps to define an entire system. As shown in Fig. 2.2, the notions of first principles and a priori concepts that lead to empiricism lead to a combination of (data) observations and formal modeling. Systems theory is based on logical positivism and integrates causal determinism to provide a formal framework for describing complex interconnected systems.

On the basis of the definition D-1 and the concept of the small world, the combination of a first concept and a plausible principle, as well as a priori concepts (see Fig. 2.2), we can establish the theoretical elements required in systems theory. This definition applies to systems that are composed of many interacting subsystems or elements that are thoroughly characterized (see workflow in Fig. 2.4).

The most impressive example of working this way is the application of statistical mechanics to discover the role of DNA molecules. This was envisaged by E. Schrödinger, who suggested that physical concepts could be applied to understand the fundamentals of biological systems. Schrödinger argued that there should be neg entropy (equivalent to information storage) to maintain a low level of entropy in living systems. Such an argument was not only an extraordinary and elegant prediction but also the trigger for further investigations that eventually led to the discovery of the DNA molecule [51].

The first principle provides the basic conceptual basis, i.e., the basic "small world", which determines the observables needed to build the entire system. This theoretical framework provides the information required for experimental design.

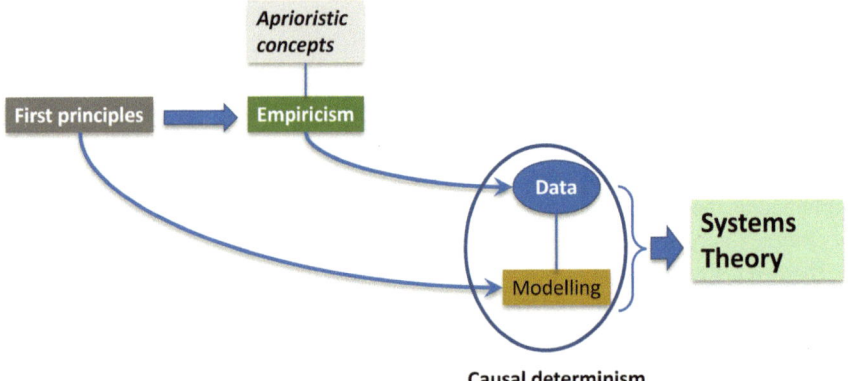

Fig. 2.4 Workflow based on an OACEM relationship, described in D-1

The systematic integration of the different elements that hypothetically belong to the system and the subsequent validation of the results lead to the definition of the final mathematical description.

The idea of reducing biology into mechanistic concepts is based on this basic canvas, which inspired the idea that biological systems, and in general any complex system, follow a set of fundamental laws, irrespective of how complex these systems are, a position that is very attractive for physicists, who tend to think about universal laws as a foundation for any phenomenon in the universe [19].

Interestingly, such an approach, which is typical of systems theory, contrasts with other generally accepted ideas that deny such mechanistic and law-based concepts for understanding complex systems [17]. For instance, J. A. Wheeler clearly postulated that life and complex systems in general are not reducible to laws and mechanisms, such as classical physics, but that a *principle of organization* should underlie all complex phenomena [68].

Notably, network and graph theory have become standard representations of complex systems and provide the conceptual basis for a unified mathematical representation of complex systems. Combined with the use of machine learning, this has enabled theoretical representations of cells, complex organisms, the brain, social and economic systems, epidemiology, etc. [8].

However, systems theory is far from being a standardized and unified theory. Systems theory does not have a simple set of equations, such as special or general relativity or electrodynamics [13]. This also means that there is no single and elegant equation to describe life.

By striving for completeness and starting from a small-world concept and using the OACEM methodology, a naked but not uniform definition of a system emerges. When trying to derive seemingly universal properties from a complex system, it is often the case that there are relevant exceptions that contradict any universal characterization of the system.

Systems theory has a toolbox of methods that help define models that represent the real world. However, such models are only proxies. They can be either perfectly validated but false, validated and correct, or even completely invalid but informative enough to provide insight into the basic properties of the system under investigation.

According to Aristotle, physical and functional causalities are inherently paradoxical. Although physical causality can be explained by reducing complexity to a clear formalism, functional causality remains a mystery.

A solution to this problem can be found by assuming that functional causality emerges from all physical processes (considering that physics is described by means of a Hamilton function, which is essentially a causal effect) combined with the context provided by the several scales of the complex system [22].

Therefore, it takes more than the holistic representation of a complex system by means of a graph (of a complex network), a method that is often used in systems theory.

2.4 Concluding Remarks

The concept of causality is fundamental in systems theory for the intuitive understanding of many physical processes and is still part of the field theories that explain how fundamental interactions take place. It is indeed part of the core of physics and systems theory and an essential element required to postulate organizational principles in complex systems.

Without a concept of causality, it is impossible to understand and control the world. Moreover, it is not possible to apply this concept without first decomposing complex systems into individual, well-characterized elements that live in so-called "small worlds", which are theoretically predictable systems. Therefore, we propose that causality and small worlds are concepts in which both are required to identify causal determinisms in a complex system that reduce the overall complexity of the system to extract the essential interacting elements and the basic bits of information of the system. Finally, causality is not only considered an underlying principle but can also be redefined as a result of the self-organization of the system and as causal emergence [66].

Nevertheless, causality is a problematic concept. Many scholars, including Bertrand Russell, have argued that causality is useless [42]. B. Russell argued that causality is not required in theories such as gravity and quantum theory.

Even systems in a critical state can develop dynamics apparently due to a single microscopic cause, such as an avalanche in a sandpile generated by a single dust grain; however, the cause of the avalanche is not the dust grain but rather the critical state of the whole sandpile. In such cases, it is impossible to define causal events but rather causal factors when a system is in a critical state[9] [35].

---

[9] This aspect is critical in understanding complex problems like climate change, where the whole system is put at a critical state due several factors. Deniers of climate change often argue that

Furthermore, causality implies that processes can be reversible, which is an inherent contradiction: while causality in Newtonian physics implies time invariance, a time arrow becomes more relevant in thermodynamics, statistics, and large complex systems where there is no time invariance. Large systems are therefore inherently irreversible and have a time direction (increase in time).

Determinism is also a problematic concept: Individual microscopic events are difficult to estimate deterministically in large complex systems. For this reason, the concept of the distribution function is introduced. These functions can also be estimated in isolation and deterministically. However, the role of different observers in real and nonisolated systems can have unpredictable effects on such distributions.

Why are causality and determinism still important? This is because systems theory is based on logical positivism [18]. This conceptual foundation implicitly establishes the conditions for a systematic understanding of a system of empirical facts, with causal determinism being a central concept that guides the relationship between empirical work and theory to derive predictive theories.

Moreover, this theoretical background is necessary not only to preserve and validate these theories but also to use these theories to control complex systems. Finally, ideally, it should be possible to derive a set of uniform equations that underlie any complex system, similar to the idea of deriving uniform equations capable of describing every field in the entire universe, a concept associated with the idea that the individual parts and the whole are deeply connected (see, for example, Heisenberg's book "Der Teil und das Ganze" [28]).

To gain control over nature and the whole universe implies that life or consciousness can be controlled and replicated, a topic that has been extensively explored in the literature, such as the story of the ancient Jewish story of the Golem, W. V. Goethe and his "Zauber Lehrling" [25], E. T. A. Hoffman and his "Die Automate" [31] to Mary Schelley and her novel Dr. Frankenstein [54].

It is a topic that has been both fascinating and scary: In all these stories, observing how the apparent control of a complex process suddenly breaks moral and natural barriers is common. In an intense and destructive manner, these creatures discover and experience life, with all its joys, pains, love and hates that they can have.

Scientific evidence implies that the governance of complex systems can be understood in a different and less romantic way. Complex systems such as organisms, entire ecosystems or even brain processes and cognition can be understood as machines that can, in principle, be hacked and controlled.

The use of such concepts has led to advances in areas such as bioengineering and medicine (e.g., neurodegenerative diseases, cf. Angarita-Rodríguez et al. [3]), as well as several initiatives in citizen science or even biohacking as a form of art and culture.[10]

---

burning fossil fuels is not the cause of the climate change; this of course is true in a strict sense, also considering that the earth is warming since approximately 100,000 years. However, this affirmation is also false, since fossil fuels are a causal factor that are accelerating a process in a system in a critical state with unforeseen consequences.

[10] https://hackteria.org/wiki/Collection_of_DIY_Biology,_Open_Source_Art_Projects.

The causal relationship between two events is not the only way to understand causality. We should consider not only causal events but also causal properties, i.e., how general system properties (and not just individual events) trigger events in observed systems. Despite these slight differences, we can infer how the relationships between interacting or correlated elements and thus how emergent properties can determine the constraints of a system lead to explainability and causal determinism.

Through science, humanity can gain a better understanding of nature and thus enhance well-being. But literature has warned us about the illusory nature of control over complex systems.

In the next chapters, we intend to explore in more detail how concepts of causality are relevant in statistical physics and systems theory and provide an overview of statistical physics and systems theory within the framework of causal theories, with the aim of deriving an axiomatic theory of complex systems. It is a theory that is very similar to field theories in physics. Both functional causality and physical causality are subject to the same principle.

Stuttgart—Dec. 2023–Aug. 2024

References

1. Adams K, Hester P, Bradley J (2013) A historical perspective of systems theory. In: IIE annual conference and expo 2013, Jan 2013, pp 4102–4109
2. Allen J-M, Barrett J, Horsman DC, Lee CM, Spekkens RW (2017) Quantum common causes and quantum causal models. Phys Rev X 7(3):031021. https://doi.org/10.1103/PhysRevX.7.031021
3. Angarita-Rodríguez A, González-Giraldo Y, Rubio-Mesa JJ, Aristizábal AF, Pinzón A, González J (2024) Control theory and systems biology: potential applications in neurodegeneration and search for therapeutic targets. Int J Mol Sci 25(1):365. https://doi.org/10.3390/ijms25010365
4. Arthur RTW (2019) Mario Bunge on causality: some key insights and their Leibnizian precedents. In: Matthews MR (ed) Mario Bunge: a centenary festschrift. Springer International Publishing, Cham, pp 185–204. https://doi.org/10.1007/978-3-030-16673-1_11
5. Barbosa de Almeida MW (2015) Structuralism. In: Wright JD (ed) International encyclopedia of the social & behavioral sciences, 2nd edn. Elsevier, Oxford, pp 626–631. https://doi.org/10.1016/B978-0-08-097086-8.12225-1
6. Bell M (2008) Hume on causation. In: Norton DF, Taylor J (eds) The Cambridge companion to Hume, 2nd edn. Cambridge companions to philosophy. Cambridge University Press, Cambridge, pp 147–176. https://doi.org/10.1017/CCOL9780521859868.005
7. Bender A, Beller S, Medin DL (2017) Causal cognition and culture. In: Waldmann MR (ed) The Oxford handbook of causal reasoning. Oxford University Press. https://doi.org/10.1093/oxfordhb/9780199399550.013.34
8. Benito RM, Cherifi C, Cherifi H, Moro E, Rocha LM, Sales-Pardo M (eds) (2022) Complex networks & their applications X: volume 1, proceedings of the tenth international conference on complex networks and their applications COMPLEX NETWORKS 2021. Studies in computational intelligence, vol 1072. Springer International Publishing, Cham. https://doi.org/10.1007/978-3-030-93409-5
9. Bertalanffy LV (1950) An outline of general system theory. Br J Philos Sci 1(2):134–165. https://www.jstor.org/stable/685808

10. Binder K (2010) Monte Carlo simulation in statistical physics: an introduction. Springer-Verlag Berlin and Heidelberg GmbH & Co. K

11. Binder DPLK (1604) A guide to Monte Carlo simulations in statistical physics by David P. Landau Kurt Binder (2014-12-15). Cambridge University Press

12. Boccaletti S, Latora V, Moreno Y, Chavez M, Hwang D-U (2006) Complex networks: structure and dynamics. Phys Rep 424(4):175–308. https://doi.org/10.1016/j.physrep.2005.10.009

13. Boogerd F, Bruggeman FJ, Hofmeyr J-HS, Westerhoff HV (eds) (2007) Systems biology: philosophical foundations, 1st edn. Elsevier Science, Amsterdam, Boston

14. Bremmer J (1999) Rationalization and disenchantment in ancient Greece: Max Weber among the Pythagoreans and Orphics? In: Buxton R (ed) From myth to reason? Studies in the development of Greek thought. Oxford University Press. https://doi.org/10.1093/oso/9780198152347.003.0004

15. Bunge M (2009) Causality and modern science, 4th revised. Transaction Publishers, New Brunswick, NJ

16. Buxton R (1999) From myth to reason? Studies in the development of Greek thought, 1st edn. Clarendon Press

17. Capra F, Luisi PL (2014) The systems view of life: a unifying vision, 1st edn. Cambridge University Press

18. Carr-Chellman DJ, Carr-Chellman A (2020) Integrating systems: the history of systems from von Bertalanffy to profound learning. TechTrends 64(5):704–709. https://doi.org/10.1007/s11528-020-00540-1

19. Cockell CS (2017) The laws of life. Phys Today 70(3):42–48. https://doi.org/10.1063/PT.3.3493

20. Cohen RS (1968) Ernst Mach: physics, perception and the philosophy of science. Synthese 18(2/3):132–170. https://www.jstor.org/stable/20114601

21. Creath R (2023) Logical empiricism. In: Zalta EN, Nodelman U (eds) The Stanford encyclopedia of philosophy. Metaphysics Research Lab, Stanford University. https://plato.stanford.edu/archives/win2023/entries/logical-empiricism/

22. Ellis GFR, Kopel J (2019) The dynamical emergence of biology from physics: branching causation via biomolecules. Front Physiol 9. https://doi.org/10.3389/fphys.2018.01966

23. Falcon A (2023) Aristotle on causality. In: Zalta EN, Nodelman U (eds) The Stanford encyclopedia of philosophy. Metaphysics Research Lab, Stanford University. https://plato.stanford.edu/archives/spr2023/entries/aristotle-causality/

24. Galam S (2012) Sociophysics: an overview of emblematic founding models. In: Galam S (ed) Sociophysics: a physicist's modeling of psycho-political phenomena. Springer US, Boston, MA, pp 93–100. https://doi.org/10.1007/978-1-4614-2032-3_5

25. Goethe JWV (2015) Johann Wolfgang von Goethe, Gesammelte Werke: Gebunden in feinem Leinen mit goldener Schmuckprägung. Anaconda Verlag, Köln

26. Goldstein H, Poole C, Safko J (2001) Classical mechanics, 3rd edn. Pearson, San Francisco, Munich

27. Harari YN (2015) Sapiens: a brief history of humankind. Harper, New York

28. Heisenberg W (2001) Der Teil und das Ganze: Gespräche im Umkreis der Atomphysik, 15th edn. Piper Taschenbuch, München Berlin Zürich

29. Hillix WA, L'Abate L (2012) The role of paradigms in science and theory construction. In: L'Abate L (ed) Paradigms in theory construction. Springer, New York, NY, pp 3–17. https://doi.org/10.1007/978-1-4614-0914-4_1

30. Hoefer C (2023) Causal determinism. In: Zalta EN, Nodelman U (eds) The Stanford encyclopedia of philosophy. Metaphysics Research Lab, Stanford University. https://plato.stanford.edu/archives/win2023/entries/determinism-causal/

31. Hoffmann ETA (2020) Die automate, 1st edn. epubli

32. Hofweber T (2023) Logic and ontology. In: Zalta EN, Nodelman U (eds) The Stanford encyclopedia of philosophy. Metaphysics Research Lab, Stanford University. https://plato.stanford.edu/archives/sum2023/entries/logic-ontology/

33. Hüllermeier E, Waegeman W (2021) Aleatoric and epistemic uncertainty in machine learning: an introduction to concepts and methods. Mach Learn 110(3):457–506. https://doi.org/10.1007/s10994-021-05946-3
34. Huy JD' (2016) Scientists trace society's myths to primordial origins. Sci Am. https://www.scientificamerican.com/article/scientists-trace-society-rsquo-s-myths-to-primordial-origins/
35. Jensen HJ (1998) Self-organized criticality: emergent complex behavior in physical and biological systems. In: Cambridge lecture notes in physics. Cambridge University Press, Cambridge. https://doi.org/10.1017/CBO9780511622717
36. Jones A, Taub L (eds) (2018) The Cambridge history of science: volume 1: ancient science, vol 1. Cambridge University Press, Cambridge. https://doi.org/10.1017/9780511980145
37. Keuth H (2015) Logical positivism and logical empiricism. In: Wright JD (ed) International encyclopedia of the social & behavioral sciences, 2nd edn. Elsevier, Oxford, pp 313–318. https://doi.org/10.1016/B978-0-08-097086-8.63047-7
38. Khondker HH (2022) Epistemology. In: Farazmand A (ed) Global encyclopedia of public administration, public policy, and governance. Springer International Publishing, Cham, pp 4054–4059. https://doi.org/10.1007/978-3-030-66252-3_411
39. Kuhn TS (1996) The structure of scientific revolutions, 3rd edn. University of Chicago Press, Chicago, IL
40. Landau LD, Lifshits EM (1991) Quantum mechanics: non-relativistic theory. Butterworth-Heinemann, Oxford, Boston
41. Leitgeb H, Carus A (2024) Rudolf Carnap. In: Zalta EN, Nodelman U (eds) The Stanford encyclopedia of philosophy. Metaphysics Research Lab, Stanford University. https://plato.stanford.edu/archives/spr2024/entries/carnap/
42. Lipkind D (1979) Russell on the notion of cause. Can J Philos 9(4):701–720. https://www.jstor.org/stable/40231124
43. Mainzer K (2010) Causality in natural, technical, and social systems. Eur Rev 18(4):433–454. https://doi.org/10.1017/S1062798710000244
44. Meyer M, Vergnaud F (2020) The rise of biohacking: tracing the emergence and evolution of DIY biology through online discussions. Technol Forecast Soc Change 160:120206. https://doi.org/10.1016/j.techfore.2020.120206
45. Murray P (1999) What is a muthos for Plato? In: Buxton R (ed) From myth to reason? Studies in the development of Greek thought. Oxford University Press. https://doi.org/10.1093/oso/9780198152347.003.0014
46. Pienaar J (2017) Causality in the quantum world. Physics 10:86. https://doi.org/10.1103/PhysRevX.7.031021
47. Rahaman F (ed) (2021) Causal structure of spacetime. In: The general theory of relativity: a mathematical approach. Cambridge University Press, Cambridge, pp 187–218. https://doi.org/10.1017/9781108837996.009
48. Rosenberg A (1993) Hume and the philosophy of science. In: Norton DF (ed) The Cambridge companion to Hume. Cambridge companions to philosophy. Cambridge University Press, Cambridge, pp 64–89. https://doi.org/10.1017/CCOL0521382734.003
49. Saunders S (2005) What is probability? In: Elitzur AC, Dolev S, Kolenda N (eds) Quo Vadis quantum mechanics? The frontiers collection. Springer, Berlin, Heidelberg, pp 209–238. https://doi.org/10.1007/3-540-26669-0_12
50. Savage LJ (1954) The foundations of statistics. Wiley, Oxford
51. Schrödinger E (1989) Was ist Leben? Die lebende Zelle mit den Augen des Physikers betrachtet. Translated by Mazurcak L, 18th edn. Piper Taschenbuch, München, Berlin, Zürich
52. Schulz M (2006) Control theory in physics and other fields of science concepts, tools, and applications. Springer, Berlin, New York. http://www.myilibrary.com?id=140214
53. Secada JEK (1990) Descartes on time and causality. Philos Rev 99(1):45–72. https://doi.org/10.2307/2185203
54. Shelley M (2023) Mary Shelley, Frankenstein. Ein Schauerroman: Das Meisterwerk der englischen Romantik gebunden in Cabra-Leder mit Silberprägung. Anaconda Verlag

55. Smith R (2024) Stone age builders had engineering savvy, finds study of 6,000-year-old monument. Nature. https://doi.org/10.1038/d41586-024-02776-w
56. Sood SK, Pooja (2024) Quantum computing review: a decade of research. IEEE Trans Eng Manage 71:6662–6676. https://doi.org/10.1109/TEM.2023.3284689
57. Steup M, Neta R (2024) Epistemology. In: Zalta EN, Nodelman U (eds) The Stanford encyclopedia of philosophy. Metaphysics Research Lab, Stanford University. https://plato.stanford.edu/archives/spr2024/entries/epistemology/
58. Tolman RC (1979) The principles of statistical mechanics, New. Dover Publications Inc., New York, NY
59. Twain M, Lauder S, McGregor W (2019) A Connecticut Yankee in King Arthur's court, mit 1 audio-CD: Helbling readers red series/level 2. Helbling
60. Uebel T (2022) Vienna circle. In: Zalta EN, Nodelman U (eds) The Stanford encyclopedia of philosophy. Metaphysics Research Lab, Stanford University. https://plato.stanford.edu/archives/fall2022/entries/vienna-circle/
61. Vasquez JA (ed) (1999) Theory construction as a paradigm-directed activity. In: The power of power politics: from classical realism to neotraditionalism. Cambridge studies in international relations. Cambridge University Press, Cambridge, pp 60–76. https://doi.org/10.1017/CBO9780511491733.007
62. Viale R (2021) The epistemic uncertainty of COVID-19: failures and successes of heuristics in clinical decision-making. Mind Soc 20(1):149. https://doi.org/10.1007/s11299-020-00262-0
63. Wang Y (2013) Reverse engineering. In: Dubitzky W, Wolkenhauer O, Cho K-H, Yokota H (eds) Encyclopedia of systems biology. Springer, New York, NY, pp 1855–1856. https://doi.org/10.1007/978-1-4419-9863-7_369
64. Whitley DS (2011) Rock art, religion, and ritual. In: Insoll T (ed) The Oxford handbook of the archaeology of ritual and religion. Oxford University Press. https://doi.org/10.1093/oxfordhb/9780199232444.013.0021
65. Wilde M, Williamson J (2016) Evidence and epistemic causality. In: Statistics and causality, pp 31–41. Wiley. https://doi.org/10.1002/9781118947074.ch2
66. Yuan B, Zhang J, Lyu A, Wu J, Wang Z, Yang M, Liu K, Mou M, Cui P (2024) Emergence and causality in complex systems: a survey of causal emergence and related quantitative studies. Entropy 26(2):108. https://doi.org/10.3390/e26020108
67. Zach R (2023) Hilbert's program. In: Zalta EN, Nodelman U (eds) The Stanford encyclopedia of philosophy. Metaphysics Research Lab, Stanford University. https://plato.stanford.edu/archives/win2023/entries/hilbert-program/
68. Zurek WH (1990) Complexity, entropy, and the physics of information: the proceedings of the 1988 workshop on complexity, entropy, and the physics of information held May–June, 1989, in Santa Fe, New Mexico. Addison-Wesley

Chapter 3
A Brief Overview of Dynamic Complex Systems and Causal Inference

Wir müssen wissen—wir werden wissen
("We must know—we will know")
D. Hilbert

Keywords Statistical physics · Control function · Emergence · Deductive modeling · Inductive modeling · Systems theory

The small-world concept and its relationship with causality, presented in the previous chapter, require constant coarsening and simplification of the microscopic properties of the interacting system's elements. Such extreme simplification is very useful for reducing the complexity of the element (and its corresponding states) into simple and quantifiable observables in establishing mathematical and computational methods to describe the essential phenomenology of the collective behavior of interconnected systems.

This chapter provides a brief overview of which deductive and inductive mathematical modeling and causal inference play a role in systems theory and how the interconnectedness of systems determines the causal properties of complex systems. In this context, statistical physics and graph theory (complex networks) are relevant.

Specifically, this chapter provides an overview of mathematical models for evaluating the observability of systems and their causal structure. If the relationship between these concepts is well understood, it is possible to derive a control theory, i.e., imagine the possibility of correctly predicting and controlling any complex system.

Despite the technical nature of this chapter, the various topics has been kept as simple as possible to provide the reader with an intuitive understanding of the role of mathematical modeling in systems theory.

J. G. Diaz Ochoa, *Complexity Measurements and Causation for Dynamic Complex Systems*, Understanding Complex Systems, https://doi.org/10.1007/978-3-031-84709-7_3

3.1   Where Are the Causal Principles in Complex Systems?

Defining clear causal rules after complex systems are dissected into simple interacting elements and developing a solid theoretical foundation for complex systems remain significant milestones in systems theory. We could have a comprehensive view of the universe on different scales as well as a deeper understanding of what life truly is, what love constitutes, and how the universe evolves if we had such a theory.

However, the assumption that complex systems can be analyzed in terms of precise causal rules is quite simplistic. Complex systems are characterized by the networking of several interacting systems across several temporal and spatial scales.

This implies that the change in the state of a given observable $O_A \to O'_A$ is not simply triggered by another state or event O_B but is the effect of several events across different scales.

To predict and model complex systems, it is essential to describe and predict these state transitions. To this end, mathematical, computational, and synthetic modeling strategies are required to put fundamental principles and axioms into a mathematical form. Statistics, particularly the principles of statistical mechanics, is a way to achieve this axiomatics. It does this by considering multiple interactions across multiple scales and using statistics to gain a better understanding of such state transitions.

The next section provides a brief overview of the ways in which deductive modeling based on statistical mechanics in physics and inductive modeling based on statistical methods are both used to extract the relationships between fundamental physical concepts and the concept of causality for the construction of deductive models.[1]

3.2   Statistical Physics and Deductive Modeling

There is a certain austerity in mathematical modeling, as mathematical tools should be the same when we describe different systems on different scales, for example, if they are a population of cells or a single cell. Sometimes we can describe the entire population by ignoring the individual elements and using an average representation of the individuals (which, in physics, corresponds to a mean-field description).

Intuitively, however, we know that the right way to describe a population is to model and sample individual elements. The process of extracting elementary and microscopic interactions from complex systems is called deductive modeling.

---

[1] The mathematical theory of complex systems and systems theory require a more extensive introduction. Readers could refer to other excellent textbooks that introduce systems theory in more detail.

3.2.1 Basic Definitions

By deductive modeling, we refer to the learning and extraction of relevant properties of each element and how it interacts with other elements and the environment it is in. Furthermore, this approach assumes that fundamental mechanisms at different scales and the interrelation of these mechanisms across scales are necessary and sufficient for understanding a system.

Therefore, from an epistemological perspective, we are extracting knowledge about the entire system by underestimating the basic elements.

All of this is based on the notion of the system's control function. Intuitively, this means that any change in the state of the element depends on a control function that limits the dynamics of the system.

In physics, this function is equivalent to the Hamiltonian function $\mathcal{H}$, which represents the energy of the system. This function depends on the microscopic interaction between interacting elements and is sampled over the entire system.

This definition is fundamental to physics because it bridges the microscopic and macroscopic properties of a system through its statistical distribution.[2] The way in which the system configuration changes depends on the whole fluctuations, which drive the change in the state of the microscopic interacting elements σ_i from a state σ_i to the next state σ_i'. This transition is constrained by $\mathcal{H}$ and its minimization (i.e., the search for the minimal values of such a function).

From the pair interaction, the constraint function $\mathcal{H}$ at the state σ_i considering the interaction with its neighbors can be defined as:

$$\mathcal{H}(\sigma_i) = \sum_{i,j} g_{ij}\sigma_i\sigma_j + C, \tag{3.1}$$

where σ_i are the microscopic states and where g_{ij} is the coupling constant between the states σ_i and σ_j.

Note that in this case, the concept of the small world is applied: this description describes how two elements, the states σ_i and σ_j, interact through the constant g_{ij}, whereas in principle, irrelevant characteristics of the interacting elements are ignored. In this case, the state σ_i will not simply trigger a change σ_j. Instead, the whole system decides whether a change in one or both states is needed.

From Eq. (3.1), the observable O of the whole system, defined as an average $\langle O \rangle$ over the whole MS, can be described in statistical physics as:

$$\langle O \rangle = \sum O(\{\sigma_i\})P_{eq}(\{\sigma_i\}), \tag{3.2}$$

which depends on the probability distribution $P_{eq}(\{\sigma_i\})$ of the state σ_i defined as

---

[2] For instance, based on Newton's equations of motion, it is possible to compute an average over several interacting elements, to obtain a macroscopic observable. This ~~one~~ is the basis of molecular dynamics.

$$P_{eq}(\{\sigma_i\}) = \frac{e^{-\frac{\mathcal{H}}{K_B T}}}{Z}, \tag{3.3}$$

where Z is the partition function, defined as

$$Z = \sum e^{-\mathcal{H}/K_B T}. \tag{3.4a}$$

This distribution depends on the system's energy $\mathcal{H}$, the Boltzmann constant K_B and the temperature T.

3.2.2 The Concept of a Statistical Ensemble

From this perspective, the concept of the ensemble is introduced, which involves a replication of the interacting elements (and their corresponding states). A micro-canonical ensemble is a collection of isolated elements in a closed receiver that can only interact with other similar elements in the receiver. Moreover, a macrocanonical ensemble enables energy and particle flows through a reservoir [49].

In this last case, the partition function looks like

$$Z = \sum e^{-(N\mu - \mathcal{H})/K_B T}, \tag{3.4b}$$

where N is the total particle number and where μ is a constant. The meaning of this set of equations is that the fate of the microscale is interconnected with the fate of the macroscale:

> The distribution of the element determines the entire control function (in this case, the energy), whereas the control function determines the possible states that each element could take.

These expressions are useful for estimating the overall distribution of the system and its dependence on control parameters such as temperature in physical systems.

This control function allows the definition of a mean-field theory, i.e., essentially the calculation of the macroscopic state of the system as a function of the microscopic distributions.

However, the exact mean-field solutions can only be computed for a few systems with special characteristics, such as the one-dimensional Ising model, i.e., for models where $\sigma_i = \begin{cases} 1 \\ 0 \end{cases}$, i.e., where the single states can only take binary states, which are ordered in a one-dimensional lattice, where $g_{ij} = -J$ is a constant across the whole lattice.[3]

---

[3] The exact analytic solution of phase transitions in spin lattices has been in part possible thanks to the replica trick, which has been developed by G. Parisi in 1979/1980. According to a personal communication with K. Binder: "*Herr Parisi hat auch viel Hunderte interessantes Paper geschrieben, über*

On the other hand, the rapid development of computational methods has enabled the explicit simulation of a system's microscopic dynamics. For this purpose, the transition probabilities of the microscopically small interacting elements can be calculated exactly, which enables an explicit simulation of the system over several scales in real time.

3.2.3 Master Equation and Monte Carlo Methods

One of the most relevant methods for computing such dynamics is the Monte Carlo method. The Monte Carlo method is based on the Master equation, which describes a balance between forward and backward state transitions in the following way:

$$\frac{d\sigma_i}{dt} = \sum_{i'} W_{i \to i'} \sigma_i - \sum_{i'} W_{i' \to i} \sigma_i. \tag{3.5}$$

The basic concept of this equation is the use of stochastics to change the microscopic states, i.e., $\sigma_i \to \sigma_i'$, to generate a change in the control function $\mathcal{H}$ of the system. Thus, the probability that the state σ_i changes into a new state σ_i' is given by a probability function that accepts only the state transition if the energy is minimized, i.e.,

$$W_{i \to i'} = \begin{cases} e^{-\frac{\mathcal{H}_{\sigma_i'} - \mathcal{H}_{\sigma_i}}{k_B T}} & \text{if } \mathcal{H}_{\sigma_i'} \leq \mathcal{H}_{\sigma_i} \\ 0 & \text{otherwise} \end{cases} \tag{3.6}$$

This leads to a Markov chain, i.e., the change in a state is determined by previous steps that are stochastically defined.

The basic idea of this equation is the definition of a dynamic that systematically searches for the minimal values of the control function $\mathcal{H}$ (Eq. 3.1). By considering the stochastics of the system, which are related to the control parameters of the system (in this case, the temperature T), it is then possible to describe how the system explores all the possible additional values of the control function $\mathcal{H}$.

A low temperature value means that the system remains trapped in local minima. As the temperature increases, the system can "jump" from one minimum value to another, creating ordered states.

For this reason, the control function $\mathcal{H}$ looks like a landscape, which can be visualized as the Alps or the Andean Mountains Range, with high peaks and deep valleys. A hiker hiking in this landscape seeks the deepest valley (or the highest peak) by exploring the various local minima (or maxima) of this function.

---

eine große Anzahl diverser Probleme, aber dass er den Nobelpreis bekam, liegt fast ausschließlich an seinen 6 allein verfassten Arbeiten 1979/1980 zur Methode der Replica-Symmetrie-Brechung".

Moreover, the transition from a maximum value to one or more available valleys in the landscape implies a break in symmetry, which is relevant for understanding phase transitions. This leads to extreme behavior in the system, which means that particle distributions cannot be described with standard and simple normal distributions.

There is a constant relationship between the fluctuations of the system and the local interactions between the system's elements. This is equivalent to the concept of fluctuation/dissipation coined by H. Hacken in Synergetics [25].

This concept is fundamental for understanding emergence (i.e., when a complex entity has properties or behaviors that its parts do not have on their own), which essentially originates from the structure of the interactions (according to Eq. (3.1)) and the whole system's fluctuations.

Essentially, the constant interrelation between the transition function and the control function $\mathcal{H}$ is that of the stochastic drive, in a natural way. A search along the different configurations of $\mathcal{H}$ eventually helps to find either the maximal or minimal values of this function (depending on the definition of the problem).

Self-organization can occur at these critical states, which are joined by scale-invariant structures, resulting in fractal-like structures. These characteristics are at the core of emergent phenomena [30].

This is the reason why many scholars refer to complex systems and emergence as something other than the sum of the parts, since the whole represents the fluctuations and optimization of the sampled function $\mathcal{H}$ that drives the microscopic dynamics.

Thus, a smart coarse-graining method can map a complex system into an optimization problem, where the "cause" of the observable $\langle O \rangle$ is the minimal state of the function $\mathcal{H}$, i.e., $\langle O \rangle [\min(H)]$

3.2.4   Concepts of Causality in Statistical Physics

From the previous definitions, we learn that it is not possible to define precise causal relationships and causal principles in the framework of statistical mechanics. Instead, the function $\mathcal{H}$ determines a field where the system's dynamics are constrained.

This confirms the discussion presented in Chap. 2 and the critique of Bertrand Russell: it is not the concept of causality but the definition of causal properties that has guided the development of modern science, from the definition of gravitational fields to the theory of systems [37].

Thus, in statistical physics as well as in field theories or even complex systems, we refer to *causal properties*, instead of causal events, to determine state transitions: *the regular patterns of a system are provided by the properties of the things that are involved in these regular patterns, and these properties can be called 'causal properties'; in statistical physics, such causal properties can be reduced to the function constraining the system's dynamics* [38].

Therefore, the concept of the field plays an important role in this context, where a field is a function that contains information about how energy is distributed in space and time, just as well as how energy is consumed and dissipated by a system [33].[4]

A causal relationship is therefore not simply a set of mechanistic connections between an element with a given state (an event) and another element (which has its own eigenstate) but rather a description of the constraints generated by the interrelation of interacting elements and a field, a concept that has been implemented, for example, using Boltzmann–Ginzburg–Landau theory to describe active matter systems[5] [10].

3.3 Examples of Deductive Modeling

As the name suggests, deductive modeling attempts to isolate and derive its fundamental mechanisms from any system to represent them mathematically. The entire system can be seen as a composition of these individual components, with complexity resulting from the combination of these simple components on an ever-increasing scale. This is known as the small-world approach.

In the next section, a brief introduction to the concepts required for deductive modeling in complex systems is provided on the basis of the statistical mechanics concepts presented in the previous sections.

3.3.1 Lattice Models

In the previous section, we assumed that the coupling between two states in the Hamiltonian function is constant, i.e., $g_{ij} = J \neq J(t)$. This is a crucial assumption because it limits the behavior and dynamics of the system to both constant coupling parameters and the topology of the lattice.

In terms of lattices, this definition represents systems with relatively simple interactions. The use of grid plots is a very simple method for obtaining a qualitative understanding of the behavior of a system, as the interactions between the elements with complex topologies are simplified into regular structures.

Nevertheless, lattices with simple interactions have complex dynamics, and it is often difficult to solve their states analytically, for example, to obtain critical parameters for systems in a critical state or in a phase transition.

---

[4] Quantum field theories have a different formalism that relies on the description of the interaction between different states through intermediary particles. For a review on the topic see Maggiore [39].

[5] See definition of field in the Feynman lectures of physics: https://www.feynmanlectures.caltech.edu/II_01.html#Ch1-S2.

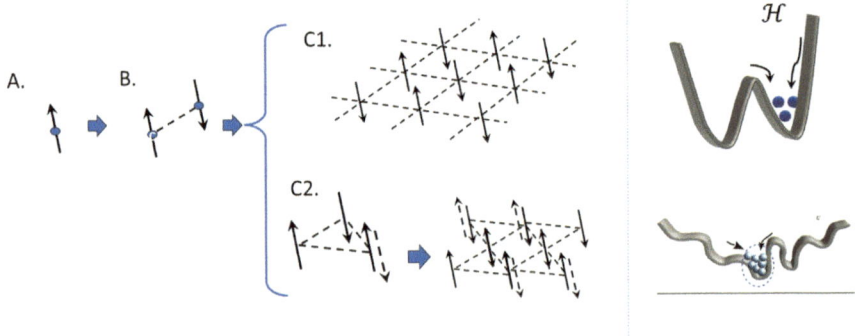

Fig. 3.1 Example of a complex system represented in a lattice. Single coarse-grained elements are assigned to simple states (**A**), and a coupling representing the interaction with a second element is defined (**B**). These interactions are then iterated in a lattice (**C**). In such cases, the pair's interaction can be simple (**C1**) or can be associated with "triplets" that might induce phenomena such as frustration (**C2**). The corresponding landscapes of the control function for the simple (**C1**) and frustrated lattices (**C2**) are illustrated on the right side

For this reason, computer experiments are necessary to gain an understanding of such systems. In Fig. 3.1, we show the basic construction of a system on the basis of systematic iteration in a lattice of pair interactions between two elements and the subsequent definition of a control function (causal property) that controls the dynamics of the system.

Note that this is an excellent example of the systematic construction of a system on the basis of abstraction and coarse graining to reduce the complexity of the interacting elements, thus defining simple concepts that are microscopically well characterized (small-world approach).

Despite this simplicity, the geometry of the couplings can increase the inherent complexity of the system and induce additional nonlinear processes [10]: whenever two elements interact in pairs, there is a reciprocal effect between their states, which can be easily characterized because it represents a microscopic causal relationship between them.

However, once a triangular motif with three elements has been introduced, it is difficult to determine the state of the third element because it does not know how to assess its state in relation to the other two elements. Simply changing grid connectivity increases system complexity (a concept known as frustrated systems; see Fig. 3.1).

Consequently, defining a model is crucial for understanding a variety of systems. Depending on the system, microscopic interacting elements are connected to each other in different ways. In addition, the complexity of the control function can vary from relatively simple (symmetrical) to more complex (asymmetrical), depending on the interaction between microscopic interactions and the control function.

Notably, in a phase transition, there is a break in the symmetry of the function. As a result, the phase transitions are constrained by the absolute minimum of the control function (this requires the computation $\min(\mathcal{H})$). Moreover, this depends not only on the coupling between the microstates but also on the coupling to external energy sources.

For example, the Ising model is an example of a representative and relatively simple model used to understand the statistical effect of individual elements that possess binary states. In such a case, the control function, which is calculated across the entire system, takes the form of a Mexican hut. This characteristic function is found not only in ferromagnetism and condensed matter but also in the description of the characteristic energy function for electroweak interactions and the Higgs mechanism [22].

The Ising model was originally defined to represent ferromagnetic materials but is also used as a principle to understand other physical systems, such as spin foams for space and time models for canonical loop quantum gravity [15], as well as biological and social systems, such as the representation of individuals in democracies choosing between two candidates [17].

The last example implies that when the complexity of a system is constrained, such as the reduction of complexity in decision-making to a binary state, it behaves like a physical system.

However, the function $\mathcal{H}$ can be very complex, particularly when this function depends on the position $\vec{x}$ of the microstates. By considering monomers with coupling functions (which are essentially energy potentials) and binding them in large chains, it is then possible to implement complex statistical models of polymers and macromolecules and represent their dynamics via Monte Carlo methods.

Using the metropolis algorithm, which is based on master Eq. (3.3), each monomer is allowed to move one step in a lattice according to the energy that defines the interaction between the monomers and the rest of the chains [34]. The representation of a complex polymer melt thus begins with the representation of individual isolated monomers; these monomers are then bound together to form long, repeating chains.

Finally, the dynamics of the melt can be calculated. In this case, the entire melt determines the position of the monomer, while the position of the monomer determines the actual energy state of the melt, which is helpful for determining the phase transitions of the melt (when the melt flows or freezes).

Note that in the previous lattice definitions, g_{ij} does not depend on time and can be defined as a constant factor, which means that the minimization of the control function, $\min(\mathcal{H})$, generally does not evolve or change as microscopic states change, i.e., there is an absolute minimum value of $\mathcal{H}$ that does not evolve along with the dynamics of the system. In addition, calculating the physical state of the system is equivalent to determining its optimal value.

However, constant couplings are not the most common case, and it can be assumed that there are cases where $g_{ij}(t)$, i.e., the couplings are dynamic and coevolve with the microstates. In this case, the essential optimization function achieves metastable optimal states, which can continue to change depending on the local transitions $\sigma_i \rightarrow \sigma_i'$.

Thus, such systems are limited not only by the parameters of the control function and the topology of the population but also by the evolution of the entire system.

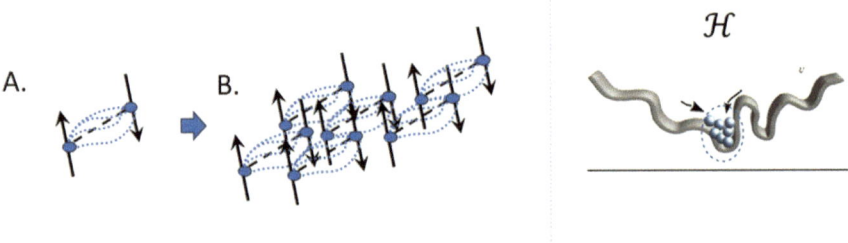

Fig. 3.2 Abstract representation of a lattice with fluctuating couplings. In contrast to Fig. 3.1, we assume that interacting elements have evolving couplings (**A**) that can then be replicated in a population in a lattice (**B**). In this case, it is also possible to define a control function (right side of the figure). However, the coupling between the interacting elements can coevolve with the whole control function

Consider, for example, populations where the coupling between individuals changes depending on their individual state (see Fig. 3.2).

An example of such a situation would be game theory on a grid. In general, game theory can be applied to populations by allowing each element to interact as if it were playing a game so that state changes correspond to the decisions made by players who have the goal of winning the game. As a result, we observe a microscopic causal relationship, since the change in the state of one element is affected by the change in the state of another interacting element (both striving for victory).

A similar abstraction can be used to represent not only rational agents but also populations of organisms and even the interaction of molecules in an abstract way.

In game theory, which is applied to populations of organisms and ecosystems, the goal is to model the cooperative behavior of the individual by defining a coupling as a matrix representing the results of a game, such as the prisoner's dilemma.

In this example, cooperation (individuals with similar states) is rewarded, whereas noncooperation (different states) is either rewarded or punished, depending on whether or not both elements defect at the same time [36].

Assuming that an individual can adopt two different states (cooperate or defect), the control function takes the following equation:

$$\mathcal{H}_i = \sum_j \sigma_i A_{ij} \sigma_j + C, \tag{3.7}$$

where A_{ij} is a matrix with different values: for binary states $\sigma_i = \begin{cases} 1 \\ 0 \end{cases}$, $A_{12} \le A_{21} <$ $A_{11} < A_{22}$. Furthermore, this score is assigned relative to each state $\mathcal{H}_i$, i.e., a transition of $\sigma_i \to \sigma_i'$, on the basis of Eq. (3.4), depends on $\mathcal{H}_i$, which is computed over the neighboring states σ_j. This equation is similar to Eq. (3.1).

Thus, instead of a constant coupling g_{ij}, we address a coupling that continuously fluctuates depending on the individual states. Thus, $\sigma_i A_{ij} \sigma_j$ has the highest score when both states σ_i and σ_j are the same (for example, where these states represent a cooperative state).

Simultaneously, different states can lead to different scores: while an element with the state σ_i can eventually obtain a higher score (for instance, when it represents noncooperation), the element with the state σ_j can obtain a low score (for instance, for cooperation).

Such variability in the scores induces a local dynamic that could also lead to more complex global dynamics as well as complex control functions $\mathcal{H}$, even if the topology where the elements are ordered in a relatively simple lattice: instead, a Mexican hut (like ferromagnetism), a complex landscape is obtained (as shown in Fig. 3.2).

Considering that the mathematical characterization of these problems is based on optimizing the characteristic function, the greatest challenge is defining methods that are able to find the absolute minimum value of the dynamic landscape without becoming stuck in local minima. The degree of mathematical complexity of such problems implies that there is no way to obtain an analytical solution for these types of mathematical models.

Thus, according to models based on game theory, the observable $\langle O \rangle$ is not only more than the sum of its parts but also a constant coevolving process between the whole and its parts.

3.3.2 Practical Consequences

As mentioned in this chapter, these concepts are helpful in understanding phase transitions and complex dynamics of off-equilibrium systems in various areas of physics, such as condensed matter and classical ferromagnetic systems (using reduced systems such as the Ising model [34]), elemental particles and field theories (Electro-Weak interactions and the Yang–Mills mechanism [22]) and quantum theories of space–time, such as loop quantum gravity and the concept of spin foams [47].

The behavior observed in physics, such as phase transitions, can also be observed in biology (e.g., coordination in flocks of birds) or in social systems (e.g., the behavior of voices in political systems that depend on control parameters and nonlinear processes) [17].

Another example, in trading and markets, is extreme behavior in the way stocks and shares are bought and resold. This extreme behavior, as well as the way prices fluctuate, can sometimes be compared to a system that is in critical condition.

To this end, mathematical models have been applied to predict prices and evaluate risk in stock markets [18]. Such methods have proven useful in recognizing that a financial crisis can be characterized very well by the dominance of short investment horizons, which goes hand in hand with the claims of the fractal market hypothesis [31].

To understand the whole theory of statistical physics, taking coevolution into account, we consider the results of a condensed matter–toy model consisting of several interacting particles with internal states that can also act either cooperatively or noncooperatively, as the model describes the adsorption of molecules suspended in a solvent on a surface so that the states of the system depend on the inherent states of each interacting element (see Fig. 3.3), so that the system is limited by the following control function:

$$\mathcal{H}_i = \sum_j \sum_{\langle k \rangle} \left(\zeta_k \hat{\Theta} \zeta_i \right)_{ij} \sigma_i \sigma_j + C \tag{3.8}$$

where ζ_i and ζ_k are the internal states of the particles and their neighbors, respectively, and where Θ is the function that modifies the interaction potential depending on the internal states.

Thus, in this equation representation, we consider that the system's couplings are fully determined by the internal element's states, $\left(\zeta_j \hat{\Theta} \zeta_i \right) = A_{ij}$.

The following is the physical significance of this condition: essentially, a surface has a potential function that attracts molecules suspended in a solvent. System dynamics can be represented with the master Eq. (3.4) for a system in a lattice and with periodic boundary conditions.

Similarly, coupled molecules can be modeled to simulate polymers adsorbed to surfaces. Usually, adsorption on surfaces is represented for molecules without interactions with solvent molecules, i.e., for systems in an ideal suspension.

However, in the implemented process, the solvents influence the adsorption of the molecules in such a way that they facilitate the binding of the molecules to the surface (cooperate) or selectively support or avoid (noncooperative/defective) by modifying the solvation barrier.

In such a system, not only the control function but also the intermolecular interactions influence the causal properties of the system [11].

Depending on the cooperative nature of the solvent, the molecules are either adsorbed more strongly on the surface (cooperative solvent), have a mixed adsorption state (tit for tat) or tend to be less adsorbent on the surface (noncooperative solvent, see Fig. 3.3).

All the systems presented in this chapter depend on the following:

  i. the coupling between the elements,
 ii. the form, or topology, of these couplings (see Sect. 3.3.1), and
iii. the density and size of the population.

Many systems can be represented in lattices. However, the geometry of the couplings also plays a relevant role in defining its control function. Not only complex interactions but also the form in which the system is organized determines the final form of the characteristic landscape of the control function.

In some systems, it can be an easy, gentle and pleasant valley. In other systems, this feature can look like an alpine landscape with multiple peaks and valleys. This will be particularly interesting when complex networks are introduced.

From all these concepts, we conclude that a careful definition of the physical boundary condition in the geometric boundary condition of the system is relevant:

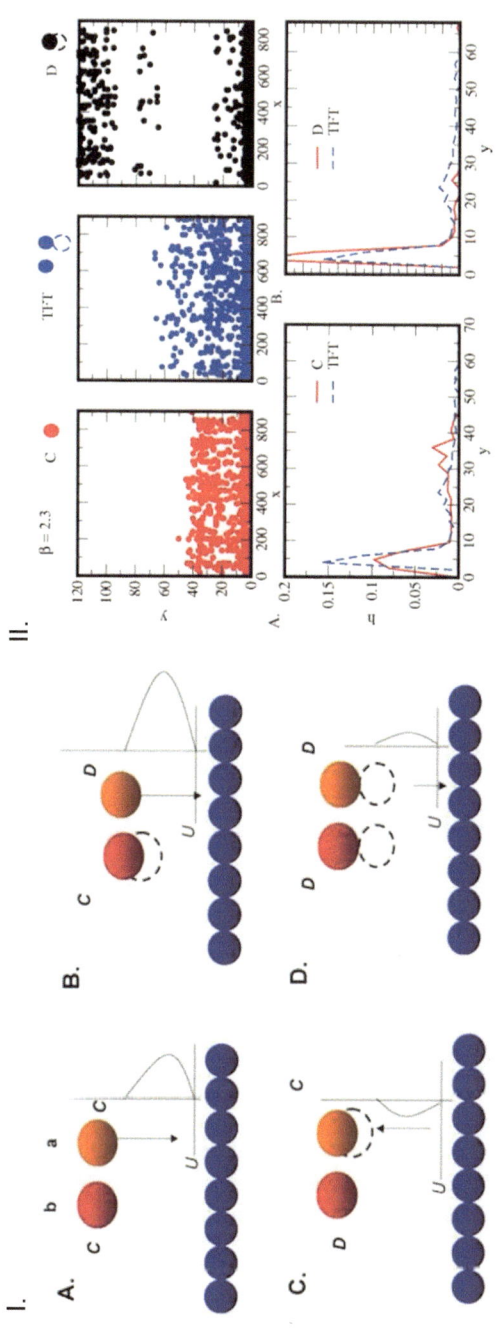

Fig. 3.3 Definition of a system representing the adsorption of molecules on a surface, considering molecules with internal states that modify the interaction potential (**I**). Thus, in A, both residues directly interact with the surface, i.e., both share the same C state. However, if a has an internal state D, then b is punished by a desolvation barrier (B), i.e.. the solvent overlaps the probe residue and target. In C, residue a is punished by a desolvation barrier if it has an internal state C and the opponent has an internal state D. In D, the solvent mediates the interaction; however, the potential is smaller than the potential associated with the direct interaction. The different solvation characteristics are represented via a game-theoretical approach, where the suspended particles and the solvent play a prisoner's role. The same experiment was repeated for the polymers (**II**). Reprinted from Diaz Ochoa [11] with permission from World Scientific

the correct definition of the physical character of the system and its topological boundary condition is encoded in the coupling factor g_{ij}.

Interestingly, this coupling resembles the concept of a metric in general relativity, where g_{ij} defines the relationship between two points in space and time; thus, from a mathematical perspective, we address both cases with a type of field theory.

We stress here that the concept "small world" does not refer to the number of neighbors of a single node in a lattice of the network but to the reduction of any complexity in the representation of a rather simple state σ_i and the coupling g_{ij} to other states.

The complexity of the interacting elements strongly decreases if one assumes that microscopic states σ_i are simple descriptions condensing the essential physical characteristics of the single interacting elements. An example of such a description would be a single binary state (e.g., Ising models) or a potential (which describes a single monomer).

Therefore, both the coarse-grained definition of the interacting elements and the geometry of the system are relevant in understanding a complex system. In other words, they are the basic causal elements that define statistical determinism.

3.4 Deductive Modeling and Complex Systems

Deductive modeling is the art of bringing all the different and seemingly incoherent pieces together to create coherent and consistent mathematical models. Since these parts are connected to each other, such a model can be understood in a similar way as mechanical watchmaking. In addition to understanding the basic laws of nature, these models usually follow the principles of economy and elegance, yet at the same time, they can be heavily distorted by the modeler. In the next sections, we introduce the concepts and theory of deductive modeling of complex systems.

3.4.1 Concepts of Emergence

The theory outlined in the previous sections is fundamental to understanding causal lines, where the whole is more than the sum of its parts. In addition, the entire system can be derived from microscopic interactions.

In this section, we explore one step further and look at systems with complex interactions. In such a case, we have a different situation than the concepts of statistical physics and phase transitions outlined in the previous section.

Here, we address complex nonlinear processes that arise from complex couplings between interacting elements.

Such interactions between the elements have led to coupling between the order parameter, for example, the gap between the measured value of the control function and its expected minimum value (see Fig. 3.1), and the control parameter.

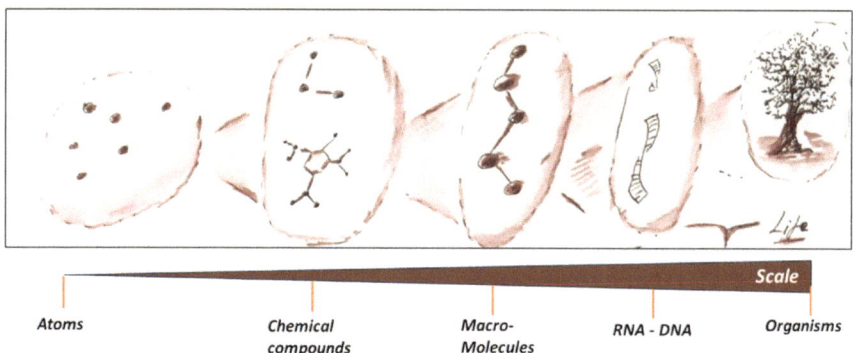

Fig. 3.4 Rough sketch illustrating how it is possible to derive large-scale systems from simple interacting elements. The concept of life can be viewed as an emergent phenomenon resulting from all the collective interactions across different scales

In self-organized systems, such coupling leads to symmetry breaks in the control function, which are not caused by changes in the control parameters but by the dynamics of the system[6] [10].

Despite the implicit concept of emergence, this theory has a clear downward causality, with each macroscopic structure depending on its collective interactions with the microscopic structures[7] [8].

The rough sketch shown in Fig. 3.4 illustrates this kind of knowledge: starting from a cluster of isolated interacting elements, such as atoms or molecules, it should be possible to identify interacting elements on larger scales. By understanding this hierarchical scaling, deriving the structure and dynamics of complex living systems from fundamental microscopic elements should be possible.

Life is therefore a product of both physical and microscopic interactions. Both individual organisms and an entire ecosystem with complex individual characteristics, such as consciousness, can be derived from such fundamental principles.

The formation of even larger structures from simple interactions implies that all systems, including biological systems, have a deterministic nature. This also implies that "free will" is unattainable [13] (the concept of "free will" has also been questioned, taking into account both the biomechanistic response of our organism and digitalism, in which digitization and information processing are used to predict various aspects of human life [26]).

Therefore, the concept of the small world is fundamental to understanding a system. Importantly, correlations between different scales can explain observed behavior, even when the system is considered as a whole. The consideration of systems and interaction networks thus provides a link to traditional philosophical discussions about reductionism versus holism [20].

---

[6] See "Die Physik lebender Prozesse": https://pro-physik.de/zeitschriften/physik-journal/2024-9/.

[7] See also this essay: https://aeon.co/essays/heres-why-so-many-physicists-are-wrong-about-free-will.

In particular, microstate interaction creates emergent fields that drive system evolution. In emergent systems, it is therefore difficult to extract clear causal interactions that constrain the evolution of the system.

Instead, these emergent fields are causal properties that drive system evolution. Notably, the concept of causal property is consistent with the basic concepts of logical positivism used to derive deterministic causality (briefly discussed in the previous chapter).

3.4.2 Biological Evolution from a Physical Perspective

The existence of collective and emergent interactions implies that there are funda-mental laws that govern any complex system. Thus, any complex system, including organisms in biology, can be described with basic equations.

By accessing such basic laws and following a strictly logical empiricism, it is not only possible to define exactly how a system can be observed but also controlled.

On the basis of this concept, some scientists have derived the so-called "equations of life", which in principle could represent the origin and evolution of any organism by limiting living processes as a constant process of increasing entropy [52]: *when a group of atoms is driven by an external source of energy (like the sun or chemical fuel) and surrounded by a heat bath (like the ocean or atmosphere), it will often gradually restructure itself to dissipate the increasingly more energy surrounding it.*

This could mean that under certain conditions, matter inexorably acquires the key physical attributes associated with life [14]. This notion, which seems to be the basis for fundamental processes in organisms, has been proposed as the equation of life:

$$\beta \langle \Delta Q \rangle_{i \to i'} + \ln\left[\frac{\pi\left(i' \to i\right)}{\pi\left(i \to i'\right)}\right] + \Delta S_{\text{int}} \geq 0 \qquad (3.9)$$

What we are showing here via Crooks' microscopic relation, however, is that the macroscopic irreversibility of a transition from an arbitrary ensemble of states $p(i|I)$ to a future ensemble $p(j|II)$ sets a stricter bound on the associated entropy production: the more irreversible the macroscopic process (i.e., the more negative $\ln\left[\frac{\pi\left(i' \to i\right)}{\pi\left(i \to i'\right)}\right]$), the more positive the minimum total entropy production must be. In other words, self-replication regulates the entropy production of the system, which should explain why replication is so relevant in biology, particularly in its evolution.

Equations such as these suggest that it is possible to assume the existence of universal physical laws underlying life as a natural phenomenon. On the basis of such a concept, it should be possible to identify any living process in any corner of the universe [9].

3.4.3 Coupled Systems

As discussed in the previous sections, complex systems are based on physical interactions, with microscopic microstates constraining the energy consumption of the system to achieve equilibrium.

This introduction shows that fundamental physical laws and fields (in this case, emergent fields) determine the microscopic dynamics.

While trying to account for all microscopic states would, in principle, be the right way to generate macroscopic systems and thus derive macroscopic behavior from the microstates. In reality, however, this method is unreliable, and coarsening becomes an essential way to derive mathematical models (coarse magnification is technically related to renormalization via methods in physics [41]).

On the other hand, it is interesting to observe that nature itself continuously generates natural or substantial coarse graining [44]. This is the basis of molecular reactions, which form the basis for mechanistic representations of biological processes such as molecular evolution, ecological networks, signal transduction, metabolism, etc.

In this way, it is possible to restore a notion of direct causality, where unambiguous causal events trigger other events. As a result, there are no fields or laws that govern the major constraints of physical dynamics but rather the way in which one state can cause a second state through a causal relationship (e.g., a chemical reaction).

These causal relationships are coupled, and their dynamics can be represented with deterministic equations, such as the coupled differential equations in the following equation:

$$\frac{dx_i}{dt} = \sum A_{ij} x_j + \lambda_i. \tag{3.10}$$

In this equation, there is again a coupling factor A_{ij} determining the relationship between the state x_i and the other system's states (with λ_i being a constant). From this equation, it is possible to define the system trajectory Γ, which is defined as the sampling of all the time series as $\Gamma = \{x_1(t), x_2(t), \ldots, x_i(t)\}$.

In Fig. 3.5, we show an example of a coupled system that describes a predator–prey system in which the individual states of each population (represented as a time series) are sampled as an overall distribution represented by the trajectory Γ.

This is a good example to illustrate how difficult it is to predict the individual states of each observable P or C; however, the overall distribution follows a rather deterministic dynamic, which illustrates that at the smallest level of a small world, there is both a coupling and a causal property that constrains the dynamics of the system.

Therefore, the concept of deterministic causality in biology and, more recently, in molecular biology has been a central element in the development of mechanistic representations of biological systems. This is accomplished by molecules that trigger or inhibit molecular processes.

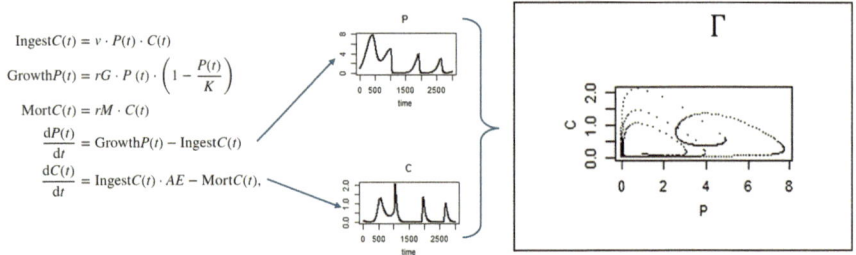

$$\text{Ingest}C(t) = v \cdot P(t) \cdot C(t)$$

$$\text{Growth}P(t) = rG \cdot P(t) \cdot \left(1 - \frac{P(t)}{K}\right)$$

$$\text{Mort}C(t) = rM \cdot C(t)$$

$$\frac{dP(t)}{dt} = \text{Growth}P(t) - \text{Ingest}C(t)$$

$$\frac{dC(t)}{dt} = \text{Ingest}C(t) \cdot AE - \text{Mort}C(t).$$

Fig. 3.5 Example of coupled equations for a population of two individuals: predators $P(t)$ and consumers $C(t)$. The trajectory of the whole population Γ, as formulated in a general form in Eq. (3.6), is thus the result of the dynamics of both $P(t)$ and $C(t)$ represented in a phase space. Depending on the specific parameters, the fluctuation can be led from one form of fluctuation to another, which is derived from the formation of attractors

However, coupled systems can become chaotic and do not have an exact analytical solution. Thus, for several coupled elements, the trajectory Γ has, in principle, a chaotic behavior with underlying patterns that depend on the system's initial state. As in Sect. 3.2, the system's dynamics are determined not only by the single interaction between the individual states but also by the system's dynamics.

The concepts we have explored in the physics of dynamical systems are therefore useful for describing such coupled systems, particularly to test the response of the system to changing parameters, as well as to test new properties such as oscillatory behavior that directly depend on the couplings of the system A_{ij} and how feedback loops influence the element states (for example, the activation of a protein in a signal transduction network).

In systems biology, this type of modeling is relevant, for example, in the representation of regulatory networks, where chemical reactions and interactions between proteins and other relevant molecules involved in biological processes can be represented as activation or inhibition processes that lead to specific biological functions such as signal transduction or the regulation of specific biological processes such as cell metabolism (see Fig. 3.6).

The dependence on the motifs and the way that the elements in such networks are connected to each other may be relevant to the limitations of the system's dynamics.

Fig. 3.6 Motif of a feed-forward reaction. In node A, the self-interaction and inhibition of node A helps maintain a basal level, i.e., a constant activation or concentration level of A

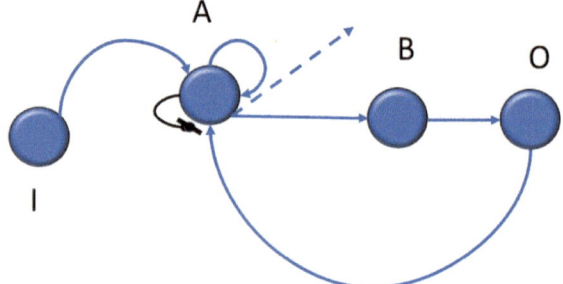

This implies that not only the interactions but also the topology of the system has a major influence on the behavior of the system. This aspect will be covered in the next section.

3.5 Inductive Modeling

In the previous sections, we presented basic concepts for deductive modeling. Intuitively, this means that we want to deduce the behavior of the entire system or population from the basic and simplest structure of the interacting elements. This type of modeling is usually defined as white-box modeling.

On the other hand, it is possible to extract patterns from recorded data from experiments and thus derive predictive models. This aims to predict events in systems. This type of modeling is called inductive modeling and is based on statistical analysis and pattern recognition of the recorded data.

Notably, such modeling is appealing because it does not require any special knowledge of the system. Instead, the availability of the data and the computing power are the decisive factors for extracting knowledge from this model.

For this reason, inductive modeling is essentially noncausal, as not only are the mechanisms not explicitly considered but also because such mathematical modeling is based on correlations that are not equivalent to causal relationships.

3.5.1 Theoretical Basis

Unlike deductive models, inductive models aim to obtain consistent data correlations that lead to stable system predictions. Mathematical correlation is usually defined as the statistical covariance between two parameters or observations. Thus, a model that links different input parameters to obtain a prediction implicitly assumes a correlation between the input parameters.

For example, linear regression is a method of calculating the covariance between two variables. This form of statistical analysis is also one of the simplest forms of machine learning. However, different forms of machine learning also allow the correlation of several parameters for nonlinear data.

Regardless of which statistical method is selected, the essential goal of inductive modeling is to map a set of parameters $\{P\}$, which can be ordered as an input vector with dimension m, into a set of output observables $\{O\}$, described as a vector with a dimension n. Accordingly,

$$f : P \mapsto O; \tag{3.11}$$

where the function f, the mapping function, can be any statistical model, such as a support vector machine (SVM), a decision tree, a neural network (NN), etc. Equation (3.11) is therefore a general mathematical representation of what essentially a machine learning model does, without considering in detail how f is actually defined.

Furthermore, simple models consist of training single f functions, whereas more complex and intelligent models automatically adjust a family of functions f_l depending on the output data by punishing low performers f_i's (which is called reinforcement learning).

Notably, models based on multiple parameters, such as neural networks and deep learning models, require a feedback loop and a simultaneous optimization of a control function that quantifies the accuracy of the model's adaptation to the data, resulting in the model behaving like a self-organizing system that acts according to the mutual interaction of the optimization of the control function (causal element) and the individual states (such interpretation is currently relevant for model explainability of neural networks; see, e.g., Diaz Ochoa et al. [12]).

Of course, machine learning models are not trained on a single vector but on a family of vectors such that the outputs O constitute a subspace of the expected solution space. These models are selected to provide efficient solutions depending on the nature and structure of the problem. Note that the family of inputs and outputs is essential: a collection of multiple observables or states σ–i can be extracted from the system.

The selection of the correlation is relevant for determining the nature of the problem, i.e., the relationship between $\{P\}$ and $\{O\}$ contains fundamental information about the nature of the system.

For example, a correlation can be used to understand the status of a population (e.g., a virus spread) in relation to its individual parameters. Similarly, the response of each individual or individuals and their likelihood of becoming infected with the virus can be influenced by modeling fine-grained correlations between molecular mechanisms and an individual's infection.

At both scales, the same mathematical methods can be used to calculate the correlations, but the model outputs are completely different.

Note that one notable aspect related to inductive modeling is that the modeler does not necessarily have to be an expert in the field represented by the model. Therefore, inductive models have an agnostic mathematical structure. Only good intuition about the potential relationship between $\{P\}$ and $\{O\}$ is needed.

As we have observed, correlation computation implicitly assumes probable inherent causal relationships between the input parameters $\{P\}$ and the observed $\{O\}$ (see Fig. 3.7).

This, however, does not mean that causal relationships can effectively be established. As a result, assuming that there is an inevitable real causal connection between $\{P\}$ and $\{O\}$ would be a fallacy.

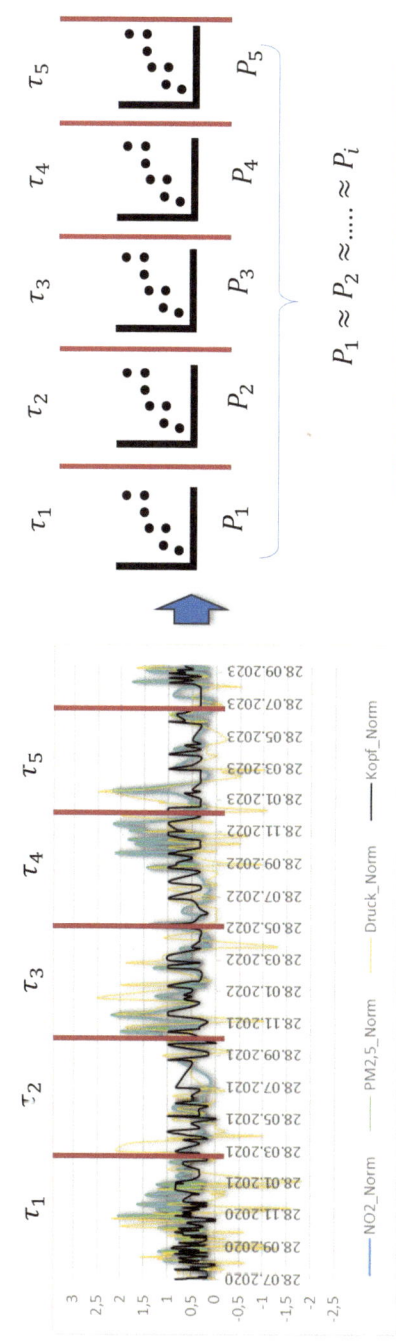

Fig. 3.7 Analysis of time series to discover causal patterns. In the time series, different time periods are extracted. From the analysis of the data in each period and the extraction of patterns, it is then possible to affirm whether a pattern in a period P_i is persistent and thus originated from a previous period P_{i-1}

3.5.2 *Inductive Modeling and Time*

The problem with this approach is that it can imply a 'post hoc, ergo propter hoc' fallacy, i.e., believing that events in the past must necessarily trigger events in the future, regardless of how causality is interpreted.

However, econometrics uses the concept of predictive causality, which differs from the strict concept of causality introduced in the previous sections [29]. The intuitive idea is to take sections of the time series, analyze their patterns, and determine if there are regularities.

If this is the case, it should be possible to predict future periods by sampling a period in a time series (see Fig. 3.7). With this idea, methods such as Granger causality on multivariate analysis were defined [28, 48]. The mathematical form of this method is as follows:

$$X(t) = \sum_{\tau=1}^{L} A_\tau X(t - \tau) + \varepsilon(t), \qquad (3.12)$$

where $X(t)$ is an d-dimensional time series, A_τ is the matrix for every τ and where $\varepsilon(t)$ is a white Gaussian random vector; this means that $X(t - \tau)$ is the cause of $X(t)$ if the factor matrix A_τ is larger than zero.

Granger causality is a classical econometric method. Machine learning also analyzes time sequences. Recurrent neural networks, in particular, use a technique to extract patterns from time series. This can be done by examining the patterns within sections derived from the time series.

However, for long time series, this method can prove to be very inefficient. These time series can also be divided into arbitrary sections, where input gates can be defined to determine which time period to parse and which to ignore. In this way, relevant long-term dependencies (the so-called LSTM method) are used to calculate predictions over the time series [19].

These are all cases where analyzing data over different time periods provides input information that can be used for predictions, assuming an inherent causal relationship between states that leads to constant patterns over time.

3.6 System Explainability and Causality in Systems Theory

To derive mathematical representations of complex systems, understanding the concepts of deductive and inductive modeling is important. In this section, we discuss the relationships between these modeling techniques and causality to derive a theoretical basis for systems theory.

3.6.1 System Explainability and Causality

From the analysis presented in the previous sections, two types of causal determinism can be identified:

1. Causal determinism is generated by causal properties such as collective interactions or fields (statistical physics or collective systems—Sects. 3.2–3.5).
2. Causal determinism is generated by causal events, which are triggered by single states (coarse-grained systems and network motifsSect. 3.6).

Instead, with the concepts of correlation and predictive causality, it is possible to establish the interrelations between elements and their corresponding states.

This concept therefore applies to deductive modeling, which is based on collective dynamics generated by the interactions between individual objects (where the collective leads to a causal property represented by the control function), or to inductive modeling (where an inherent causal property controls the linkage between parameters and observables).

In this context, causal events can be considered coarse-grained states of microscopic elements, which are also determined by causal elements. This is perhaps why a classical understanding of causality is not widely used (and/or not useful) in understanding complex systems. In general, understanding a causal property, such as the control function, is often fundamental to modeling complex systems.

In a multiscale and emergent system, how microscopic elements are interrelated through coupling g_{ij} and how critical parameters are set to generate a change in the system's observability $\langle O \rangle$.

The Markov chain is the basic concept for this type of causality in statistics, where subsequent changes in microstates and their responses to fluctuations in the system lead to changes in the entire system. In such a scenario, causality cannot be interpreted as elements or states that trigger subsequent events.

Although scholars such as B. Russell have denied causality's role in physics, it is apparent that there is a notion of deterministic causality rooted in "causal properties" in any system [16, 37]. This can be done either by accounting for direct events or by recognizing that a system's response to a parameter change is the result of an interaction across multiple scales, which forms the basis for mathematical modeling.

Therefore, interaction and the recognition of strong connections are extremely relevant. This microscopic causality can be identified in graph models such as polymer models (with single monomers interacting through potentials, similar to spheres connected by springs) or in general network models [34].

3.6.2 Role of Graph Theory in System Explainability

Graph theory and complex networks as subfields (a concept often used in physics) have become a relevant mathematical basis for describing correlated elements that are interconnected in systems with complex topologies and have become the theoretical foundation of systems theory.

From their biological structures and tissues to physics and condensed matter, complex networks are not just a theoretical construct.

However, networks are also used as allegories for abstract invisible networks, such as social interactions. In addition, complex topologies of interconnected elements can be reconstructed with the help of causal inference from time series [42].

The concept of complex networks summarizes all the aspects discussed in this chapter: they are the product of fundamental physical interactions. In addition, they are a practical way to illustrate the interrelationship between interacting elements, which are coarse-grained representations of aggregated microscopic states (which in turn can be described as subnetworks).

Complex networks are also a practical way to relate multiple disparate elements across multiple scales; they are the way to create holistic descriptions of complex systems. Finally, they define complex control functions that determine complex effects. In short, complex networks are topological representations of a field theory that constrains the dynamics of a complex system [40].

More relevant is the fact that the topology of the network, similar to spin glasses or game theory over lattices, plays a relevant role in the dynamics and function of the system. In small networks, motifs can provide complex dynamics, which in turn can explain certain biological functions.

From the topology, it can be deduced whether all the elements have a similar influence on their neighbors, depending on the edge distribution, such as an even distribution (all the nodes have the same average number of edges) or small-world networks, with high clustering and short distances; this is similar to the concept of "my friends' friends are probably also my friends" (this concept should not be confused with the concept of the "small world" described in this book).

Another common topology is the scale-free network, where edge distribution follows the power law when key nodes are central and have more connections than peripheral nodes with a low connection density (for example, networks with a scale-free edge distribution).

In general, it is assumed that the structure of a network determines whether the network is robust against damage, i.e., the network topology as a whole is determined by its robust function. A scale-free network works much more efficiently and robustly when edges are systematically separated than when they are uniformly distributed [3, 4].

In systems theory, there is a notion of organizing principles [21], which does not automatically imply that complex systems in general are governed by laws [53]. On the other hand, the concept of laws, such as the scale-free distribution principle, was coined a law of complex networks (graphs) describing complex systems [4] (Fig. 3.8).

For example, it has been suggested that network proteins are scale-free, i.e., that the edge distribution on a node follows a constant self-replicating structure, which can be described as the power law $P(k) \sim k^{-\gamma}$, where k is the number of edges connected to a node, and γ is a constant.

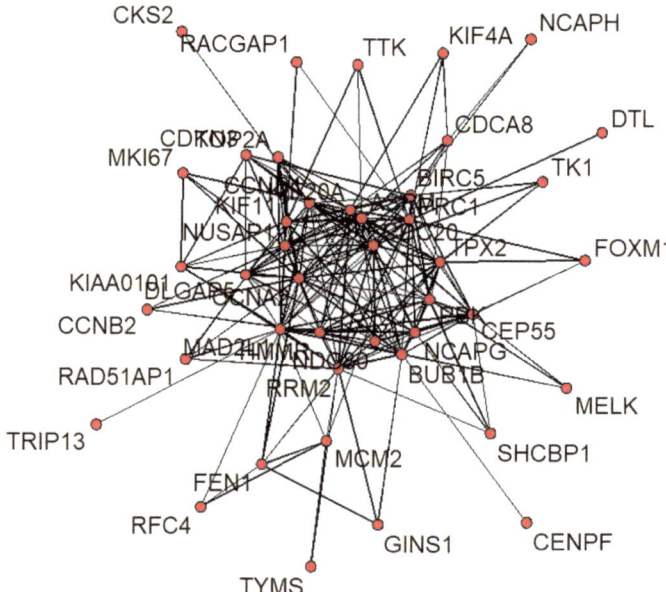

Fig. 3.8 An example of a protein network that follows the scale-free principle (few nodes with many edges and several nodes with few edges)

This should fulfill an important biological function: if the nodes in the network are removed, the network can continue to function, for example, in the transmission of information between nodes; just by removing a central node, it can actually affect the functioning of the network.

We find here that we are dealing with both physical and functional causality in networks (1st and 4th causal categories, according to Aristotle—Chap. 2).

As shown in the previous sections, individual network motifs are also a basic concept for explaining complex emergent behavior, as shown in the feed-forward model (see Fig. 3.6). In addition, multiscaling can be detected when groups of nodes are grouped together to define an average node (midfield).

These coarse grains reduce the complexity of the local network while retaining key functional properties (see Fig. 3.9).

Despite networks being rigid constructs, they can also evolve. This implies that edges are not simply time-invariant elements; rather, they can grow and evolve as part of the system dynamics.

This aspect is of interest from a theoretical point of view, considering that the elements in the network can influence the structure and constraints of the network while guiding the local dynamics of each element [23].

Such a concept is important, e.g., using game theory. However, statistical analysis via machine learning methods is also used to discover evolving features and causal patterns from data that lead to the evolution of complex networks of real-world data [42].

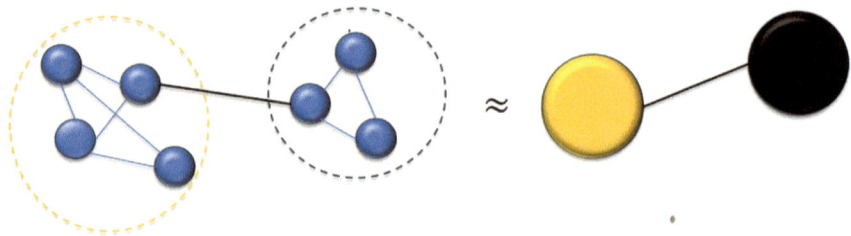

Fig. 3.9 Multiscaling and networks: two network clusters can be defined as an average (coarse-grained representation of the cluster). This final representation led to a reduced network between two nodes that ignored the microscopic characteristics of the clusters, i.e., the explicit representation of the subnetworks

Systems theory and molecular biology are closely linked, with knowledge of biological mechanisms generated by the dual strategies of decomposition and localization of constituents and molecular operations, such as protein–protein interactions (which are involved in cell regulation) or the reactions involved in metabolic reactions [1, 20].

Furthermore, network modeling and computational simulations in systems biology allow the study of larger integrated systems. These simulations provide strategies for recomposing findings in larger systems. Systems in this context, therefore, refer to large networks of integrated components exhibiting nonlinear dynamics.

Therefore, knowledge of complex networks is not only the basis of some kind of fundamental theory of systems theory but also the basis for the control and design of biology from biotechnology to "biohacking", which is a trend often led by amateurs outside academic institutions, who work on technologies that aim to manipulate or influence the biology of their own bodies [24]—for example, Bryan Johnson beating all odds to live his longest life [2].

Therefore, the control of biological pathways has become not only an academic exercise but also a lifestyle and inspiration for art and culture, with artists developing and implanting artificial tissue to modify and even enhance the human body.

Is it possible that graph theory, when applied to complex systems, has changed the way we perceive and interpret the world? Maybe yes: While calculus was the language of science and the natural sciences in past centuries, the concepts of complex networks (together with topology and differential topology) have become the language for understanding and interpreting complex systems in general.

3.6.3   Genetics, Causality, and Bioinformatics

Previously, we discussed the importance of causal properties for understanding dynamic processes. Given the many interacting elements, it is impractical (if not a fallacy) to try to discover and characterize the system in terms of causal events.

As a result, system dynamics are guided by constraints resulting from collective interactions. Such concepts are crucial for both inductive and deductive modeling.

In systems biology, however, there is still a persistent idea of causality and a small world in which individual elements trigger complex dynamics. In fact, the genome is thought to be the fundamental cause of every biological process.

We can understand the function of DNA as a "guideline" for the reproduction of macromolecules. As a result, any transcription of DNA into mRNA or RNA is similar to deciphering the guidelines used to encode these molecules. Like with any policy, the decoding process is robust and replicable. However, as with any directive, its interpretation depends on the particular conditions in which it is applied, and there may be variations (see infographic in Fig. 3.10).

Genes determine the formation of proteins, as they are their blueprint. However, a building instruction is not a cause in a narrower sense and is subject to constant change and reinterpretation (which is equivalent to constant mutation). Instead, the final form and function of the molecules created by these instructions are the ultimate cause.

For this reason, a complex disease such as cancer does not have a single causal trigger, such as a genetic factor. Instead, a combination of several factors (such as the environment, habits, and genetics) leads to this disease [35].

Causal properties remain relevant for complex biological systems. However, to reproduce complex structures with specific causal factors that lead to the function of a given organism, it is necessary to have instructions to construct the structure of the molecules connected to each other (see Fig. 3.10). Changes in function and responses also affect the interactome and ultimately the DNA code. In such cases, the system mutates and takes on new functions or adapts to new conditions.

In such systems, we describe not only the network dynamics but also the information required to reproduce the structure of the network. As a result, information, along with physical constraints, plays a crucial role in defining the function and structure of the system.

Accordingly, genes are coupled with other factors defining the complex system, but they are not the ultimate causal element in biology [45].

Extracting and interpreting these instructions has always been a challenging problem when applying mathematical modeling. For example, hypotheses were generated and tested by analyzing the response variables of the system (deductive modeling).

However, this is a rather inefficient way to decipher these natural guidelines, not only because each hypothesis needs to be tested and validated separately in different experiments but also because hypothesis generation can be influenced by other factors.

Instead, the data-driven hypothesis (inductive modeling) shows that this is a much more efficient way to understand how genes control the formation of macromolecules that are essential for biological processes. Interestingly, conventional fine-tuning of models is being replaced by novel paradigms in which context plays a relevant role.

To this end, language models are able to extract information taking into account the context to solve various tasks, such as pattern extraction from sequence data,

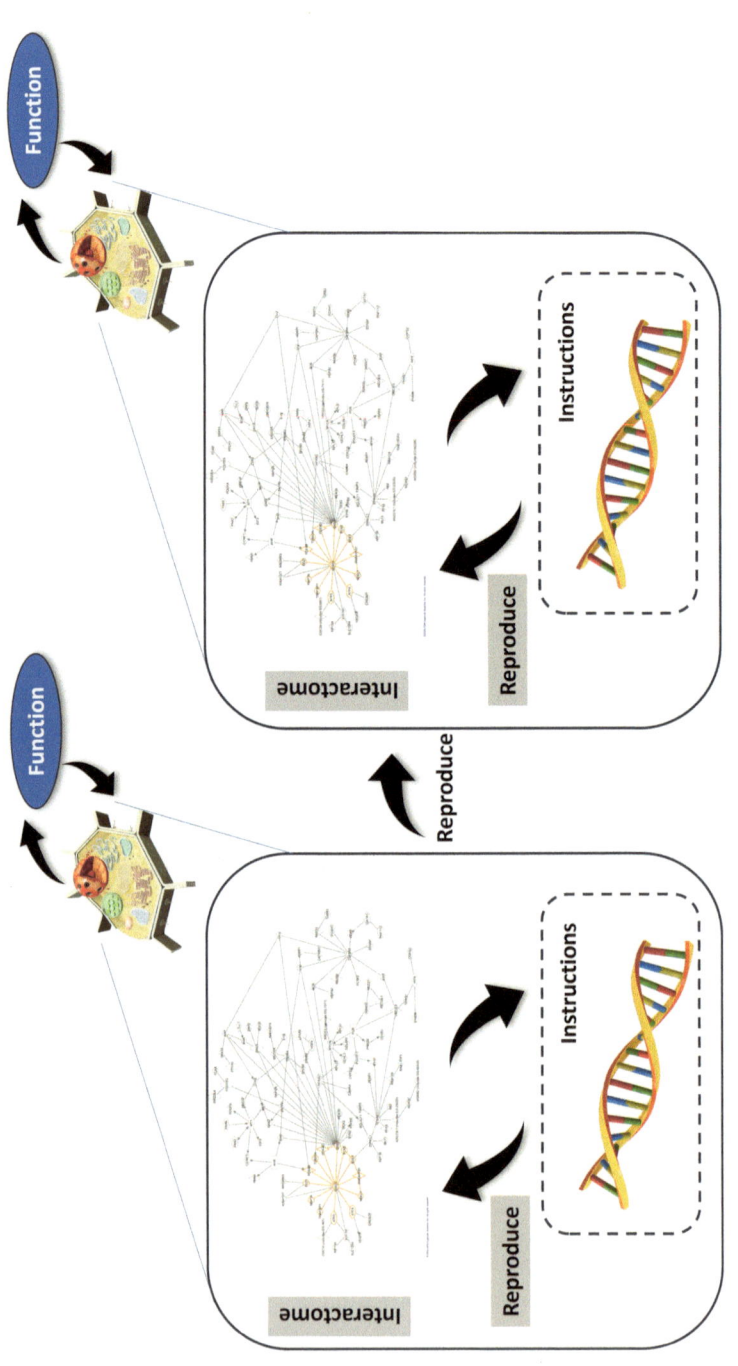

Fig. 3.10 In complex biological systems, there is not only dynamic but also information processing. Despite the fact that the interactome is related to the system's function, the information encoded in DNA molecules is responsible for replicating the interactome and, therefore, ensuring the system's functionality (interactome figure from Hennah and Porteous [27])

to solve tasks such as promoter extraction from sequence data (taking into account that the promoter region is close to the transcription start sites and regulates the transcriptional initiation of the gene by controlling RNA polymerase binding [46]).

3.7   Concluding Remarks

In this chapter, we present an overview of the theoretical aspects of statistical physics and complex systems. We have discussed how classical causal concepts have marginal utility, particularly in the representation of coarse-grained motifs. Instead, the concept of a causal property is useful in understanding how single interacting elements can lead to a collective dynamic.

The way in which individual elements interact with the control function allows us to establish a similarity with concepts of general relativity and gravitational theory: according to this theory, the object instructs space how it should be curved, and space indicates what the object's trajectory should be. Similarly, in complex systems, the interacting elements tell us how to define the control function for the entire system.

In addition, the control function (causal property) restricts the dynamics of individual elements and instructs them how to define their individual states (see infographic, Fig. 3.11).

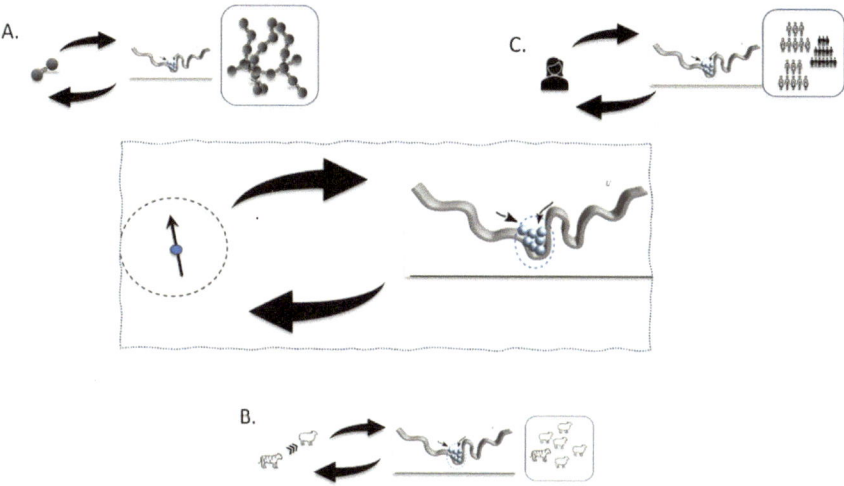

Fig. 3.11 A system's interacting elements define the control function; at the same time, the control function constrains the dynamics of the individual elements. This concept is applied to physical systems such as polymers (**A**), biological systems, population dynamics, and ecologics such as prey–predator systems (**B**), or social systems such as electoral systems with individuals deciding between two candidates (**C**)

According to this principle, establishing a causal origin for a complex event is not trivial. For example, in winter, there is usually a high occurrence of avalanches in the Alps, and under critical conditions, a tiny snow crystal can trigger a large and devastating avalanche.

However, this is not possible if the collective conditions in the snowfield (temperature fluctuations, fresh snow, high slopes, etc.) are not in a critical state. Thus, it is not the single crystal (microscale or single state) or the entire snowfield (macroscale, causal property) that triggers the snow avalanche but a combination of both.

This example can be extrapolated and used to understand how causal events in other complex systems, from biomedicine (e.g., triggering cancer) to weather conditions and climate change[8] or even in societies (for instance, where democratic societies collapse into a dictatorial state).

This principle can be used to model and explain different systems by extracting their basic properties. Depending on the causal property, the landscape can be more or less complex. For example, systems consisting of interacting elements that possess binary states that are in a critical state tend to organize themselves on different scales.

In physical systems, this is associated with phase transitions. Interestingly, phase transitions can also occur in nonphysical systems such as societies [51]. The systematic cognitive reduction and limitation of an individual's cognitive abilities can drive societies to behave like a physical system; in such cases, it is important to observe that societies can be put into a critical state when the complex behavioral characteristics of its members of society are excessively reduced so that it is possible to induce a kind of polarization between groups at high social temperatures, thus inducing phase transition-like conditions, as if the system were a physical system, which consists of particles instead of individuals.

Systems in a critical state effectively have some kind of universal behavior depending on the interaction structure and topology of the element population, which is based on complex networks that affect system dynamics and interaction structure.

Networks with a scale-free distribution are closely related to fractal structures, i.e., they represent the self-reproduction of similar structures on different scales. These types of structures are observed in a variety of systems ranging from galaxies to broccoli, suggesting that some type of universal behavior, and therefore a fundamental law, may exist [3].

In addition, networks with this type of edge distribution have the added benefit of providing a certain level of robustness, meaning that they remain connected even when nodes are systematically separated.

The theories presented in this chapter, as well as the principle outlined in Fig. 3.9, suggest that there are real basic principles that guide complex systems. In addition, it can be assumed that all these principles are based on causal properties.

Such principles are not axioms (i.e., laws) in the strict sense of the word but fundamental elements that allow the folding of complex systems.

---

[8] The misunderstanding is a tool used by opportunists, such as climate change deniers, to obscure complex interrelationships. As a result, they affirm that human activity does not play an active role in climate change.

A number of biological systems (such as molecular structures and physiological neural networks) as well as social systems (such as social networks) possess an additional "natural" network/graph-like structure that describes the interaction between nodes/vertex through the edges and links.

There are also some systems that could be efficiently described with a graph model but did not originally have this structure. A prominent example is the representation of molecular interactions in cells and tissues.

The second prominent example is the injection of medical knowledge in the form of knowledge graphs or the structuring of the Electronic Health Records (HER) itself, which is usually stored as structured databases, such as graphs [50].

Finally, information plays a relevant role in defining the basic principles of some complex systems. These systems, like biological systems, have developed the ability to encode instructions to guide the construction of complex networks that provide the most efficient functions, such as metabolism and regulation.

Therefore, information plays an essential role in this context, as it provides a way to encode "guidelines", enabling robust replication of a system [53]. This aspect differs from classical field theory: instead of having only causal properties, information ensures that certain causal properties are reproduced in other new systems, so these new systems do not have to start from scratch to define and constrain their dynamics.

However, the dream of a comprehensive and complete and general formalism that combines all the principles of complex systems persists. Why are these basic and perhaps universal principles of complex systems not fully accepted? More specifically, why is there no general theory about how adaptive functionality arises from large collections of individual decentralized components (such as ant colonies)? [43].

In general, there is no unified view of what constitutes a systems approach [6, 7]. Furthermore, systems biology comprises different research practices with different ties to molecular biology, genomics, and nonbiological disciplines [32].

Finally, many studies have reported that principles such as freedom of scale in networks as a form of rationale and robustness are rare and not well supported by empirical data. This means that the freedom of scale is not a law, as many researchers have suspected [4, 5]. Therefore, a revision of causal determinism is still needed, which is a topic that will be discussed in the next chapter.

[Stuttgart, Allgäu and Domaso—2024]

References

1. Alon U (2019) An introduction to systems biology: design principles of biological circuits, 2nd edn. Chapman and Hall/CRC, Boca Raton
2. Angeles CA/Los, and Photographs by Philip Cheung for TIME (2023) The man who thinks he can live forever. TIME. 20 Sept 2023. https://time.com/6315607/bryan-johnsons-quest-for-immortality/
3. Barabási A-L (2009) Scale-free networks: a decade and beyond. Science 325(5939):412–413. https://doi.org/10.1126/science.1173299

4. Barabási A-L (2017) The elegant law that governs us all. Science 357(6347):138. https://doi. org/10.1126/science.aan4040
5. Broido AD, Clauset A (2019) Scale-free networks are rare. Nat Commun 10(1):1017. https:// doi.org/10.1038/s41467-019-08746-5
6. Calvert J (2010) Systems biology, interdisciplinarity and disciplinary identity. In: Collaboration in the new life sciences. Routledge
7. Calvert J, Fujimura JH (2009) Calculating life? A sociological perspective on systems biology. EMBO Rep 10(S1):S46–S49. https://doi.org/10.1038/embor.2009.151
8. Capra F, Luisi PL (2014) The systems view of life: a unifying vision, 1st edn. Cambridge University Press
9. Cockell CS (2017) The laws of life. Phys Today 70(3):42–48. https://doi.org/10.1063/PT.3. 3493
10. Denk J, Frey E (2020) Pattern-induced local symmetry breaking in active-matter systems. Proc Natl Acad Sci 117(50):31623–31630. https://doi.org/10.1073/pnas.2010302117
11. Diaz Ochoa JG (2009) A model for solvent mediated interactions between molecules and surfaces. Int J Mod Phys C 20(05):747. https://doi.org/10.1142/S0129183109013960
12. Diaz Ochoa JG, Maier L, Csiszár O (2021) Bayesian logical neural networks for human-centered applications in medicine. https://doi.org/10.1101/2021.11.15.21266351
13. Ellis GFR (2023) Efficient, formal, material, and final causes in biology and technology. Entropy 25(9):1301. https://doi.org/10.3390/e25091301
14. England JL (2015) Dissipative adaptation in driven self-assembly. Nat Nano 10(11):919–923. https://doi.org/10.1038/nnano.2015.250
15. Engle JS (2014) Spin foams. In: Ashtekar A, Petkov V (eds) Springer handbook of spacetime. Springer handbooks. Springer, Berlin, Heidelberg, pp 783–807. https://doi.org/10.1007/978-3-642-41992-8_38
16. Frisch M (2023) Causation in physics. In: Zalta EN, Nodelman U (eds) The Stanford ency-clopedia of philosophy. Metaphysics Research Lab, Stanford University. https://plato.stanford. edu/archives/win2023/entries/causation-physics/
17. Galam S (2012) Sociophysics: an overview of emblematic founding models. In: Galam S (ed) Sociophysics: a physicist's modeling of psycho-political phenomena. Springer US, Boston, MA, pp 93–100. https://doi.org/10.1007/978-1-4614-2032-3_5
18. Gallo C (2010) Mathematical models of financial markets. In: Capecchi V, Buscema M, Contucci P, D'Amore B (eds) Applications of mathematics in models, artificial neural networks and arts: mathematics and society. Springer Netherlands, Dordrecht, pp 123–130. https://doi. org/10.1007/978-90-481-8581-8_6
19. Gers FA, Schmidhuber J, Cummins F (2000) Learning to forget: continual prediction with LSTM. Neural Comput 12(10):2451–2471. https://doi.org/10.1162/089976600300015015
20. Green S (2017) Philosophy of systems and synthetic biology, June 2017. https://plato.stanford. edu/Archives/win2021/entries/systems-synthetic-biology/#NetwApprSystBiol
21. Green S, Wolkenhauer O (2013) Tracing organizing principles: learning from the history of systems biology. Hist Philos Life Sci 35(4):553–576. https://www.jstor.org/stable/43862214
22. Griffiths D (2008) Introduction to elementary particles, 2 überarbeitete. Wiley-VCH, Weinheim
23. Gross T, Blasius B (2007) Adaptive coevolutionary networks: a review. J R Soc Interface 5(20):259–271. https://doi.org/10.1098/rsif.2007.1229
24. Gruber K (2019) Biohackers. EMBO Rep 20(6):e48397. https://doi.org/10.15252/embr.201 948397
25. Haken H (2012) Synergetics: an introduction, 3rd edn. Softcover reprint of the original 3rd edn. 1983 edition. Springer
26. Harari YN (2016) Yuval Noah Harari on big data, Google and the end of free will. Financial Times, 26 Aug 2016. Sec. FT Magazine. https://www.ft.com/content/50bb4830-6a4c-11e6-ae5b-a7cc5dd5a28c
27. Hennah W, Porteous D (2009) The DISC1 pathway modulates expression of neurodevelop-mental, synaptogenic and sensory perception genes. PLoS ONE 4(3):e4906. https://doi.org/10. 1371/journal.pone.0004906

28. Hlaváčková-Schindler K, Naumova V, Pereverzyev Jr S (2016) Granger causality for ill-posed problems: ideas, methods, and application in life sciences. In: Statistics and causality. Wiley, pp 249–276. https://doi.org/10.1002/9781118947074.ch11
29. Imbens GW, Rubin DB (2015) Causal inference for statistics, social, and biomedical sciences: an introduction. Cambridge University Press, Cambridge. https://doi.org/10.1017/CBO9781139025751
30. Jensen HJ (2023) Complexity science: the study of emergence, New. Cambridge University Press, Cambridge, United Kingdom, New York, NY
31. Kristoufek L (2013) Fractal markets hypothesis and the global financial crisis: wavelet power evidence. Sci Rep 3(1):2857. https://doi.org/10.1038/srep02857
32. Krohs U, Callebaut W (2007) Data without models merging with models without data. In: Boogerd FC, Bruggeman FJ, Hofmeyr J-HS, Westerhoff HV (eds) Systems biology. Elsevier, Amsterdam, pp 181–213. https://doi.org/10.1016/B978-044452085-2/50011-5
33. Landau LD (1980) The classical theory of fields, 4th edn. Pergamon, Oxford, New York
34. Landau DP, Binder K (2005) A guide to Monte Carlo simulations in statistical physics. Cambridge University Press
35. Ledford H (2024) Why are so many young people getting cancer? What the data say. Nature 627(8003):258–260. https://doi.org/10.1038/d41586-024-00720-6
36. Lewis HM, Dumbrell AJ (2013) Evolutionary games of cooperation: insights through integration of theory and data. Ecol Complex 16:20–30. https://doi.org/10.1016/j.ecocom.2013.02.007
37. Lipkind D (1979) Russell on the notion of cause. Can J Philos 9(4):701–720. https://www.jstor.org/stable/40231124
38. Loke A (2022) Causation and laws of nature. In: Loke A (ed) The teleological and Kalam cosmological arguments revisited. Palgrave frontiers in philosophy of religion. Springer International Publishing, Cham, pp 37–70. https://doi.org/10.1007/978-3-030-94403-2_2
39. Maggiore M (2005) A modern introduction to quantum field theory, Illustrated edn. Oxford University Press, Oxford, New York
40. Mata ASD (2020) Complex networks: a mini-review. Braz J Phys 50(5):658–672. https://doi.org/10.1007/s13538-020-00772-9
41. McComb WD (2003) Renormalization methods: a guide for beginners. Oxford University Press. https://doi.org/10.1093/oso/9780198506942.001.0001
42. Mei P, Zhao YH (2024) Dynamic network link prediction with node representation learning from graph convolutional networks. Sci Rep 14(1):538. https://doi.org/10.1038/s41598-023-50977-6
43. Mitchell M (2010) Biological computation, Sept 2010. Computer Science Faculty Publications and Presentations. https://pdxscholar.library.pdx.edu/compsci_fac/2
44. Morita K (2023) A fine-grained distinction of coarse graining. Eur J Philos Sci 13(1):12. https://doi.org/10.1007/s13194-023-00513-0
45. Noble D (2006) The music of life: biology beyond genes. Oxford University Press, USA
46. Oubounyt M, Louadi Z, Tayara H, Chong KT (2019) DeePromoter: robust promoter predictor using deep learning. Front Genet 10. https://doi.org/10.3389/fgene.2019.00286
47. Rovelli C (2018) The order of time. Allen Lane, London
48. Runge J, Gerhardus A, Varando G, Eyring V, Camps-Valls G (2023) Causal inference for time series. Nat Rev Earth Environ 4(7):487–505. https://doi.org/10.1038/s43017-023-00431-y
49. Tolman RC (1979) The principles of statistical mechanics, New. Dover Publications Inc., New York, NY
50. Tong C, Rocheteau E, Veličković P, Lane N, Liò P (2022) Predicting patient outcomes with graph representation learning. In: Shaban-Nejad A, Michalowski M, Bianco S (eds) AI for disease surveillance and pandemic intelligence: intelligent disease detection in action. Studies in computational intelligence. Springer International Publishing, Cham, pp 281–293. https://doi.org/10.1007/978-3-030-93080-6_20
51. Tsarev D, Trofimova A, Alodjants A, Khrennikov A (2019) Phase transitions, collective emotions and decision-making problem in heterogeneous social systems. Sci Rep 9:18039. https://doi.org/10.1038/s41598-019-54296-7

52. Wolchover N (2014) A new physics theory of life. Quanta Magazine. 22 Jan 2014. https://www.quantamagazine.org/a-new-thermodynamics-theory-of-the-origin-of-life-20140122/
53. Zurek WH (1990) Complexity, entropy, and the physics of information: the proceedings of the 1988 workshop on complexity, entropy, and the physics of information held May–June, 1989, in Santa Fe, New Mexico. Addison-Wesley

Chapter 4
Mathematical Basis: Elastic States and Complex Dynamics

Keywords Topological multiscaling · Elastic states · Distortion module · Dynamic systems · Incompleteness

As discussed in the previous chapter, the small-world approach is a systematic method for decomposing and generating knowledge from complex systems. By simplifying the complexity and neglecting the inherent structure of interacting elements, the essential features are isolated and mapped to mathematical representations to develop inductive or deductive models. We have also shown that this can be a theoretical way to define a unified theory of complex systems similar to field theory.

However, we have also learned that these concepts have limitations and that real-world complex systems still resist this type of reductionism.

In this chapter, we revisit reductionism in multiscale systems by considering the element's context. We also consider that interacting elements can assume elastic states [13] to determine their context, a notion that can contribute to molecular evolution [14].

This implies that the coarse-grained interaction of the elements that results in mathematical representations of entire systems is valid only under the conditions under which the coarse-grained representation is performed, which indicates that the representation is not universal.

This mathematical and conceptual framework is essential for understanding possible mathematical constraints in causal inference: while a small-world concept (context-agnostic) leads to complete mathematical representations, real, irreducible complex systems may not have complete representations from a mathematical point of view; i.e., it is not possible to derive clear axioms, particularly to establish causal relationships or causal properties.

This implies that a different concept of diversity and autonomy might lead to persistent data errors. We have investigated how persistent topologies and concepts inspired by integrated information theory can be used in causal inference.

4.1   Irreparable Incompleteness in Complex Systems

Are complex systems and the entire universe deterministic? If that is the case, then there are mechanisms that can be accessed by anyone in every corner of the universe; as such, there is no place for free will in a deterministic universe. However, are there limits to such a concept? Let us take a detailed look at the arguments.

In the previous chapter, two main causal concepts, **causal properties** (for instance, physical laws) and **causal events** (such as signaling pathways in cells), were defined as fundamental principles in defining deterministic systems [20]. In this context, the concept of determinism refers to the fact that a system's state in time t_2 depends on its previous state $t_1 < t_2$.

Individual states are the observable properties of interacting elements (either particles or agents). The couplings between the states are the source of causal properties that limit the dynamics of the system. This is where the concept of the small world comes into play: starting from the knowledge of each microscopic and well-defined state, it is then possible to identify the main constraints and extract the dynamics of the system.

As a result, causal properties determine the trajectory of the system. This trajectory is defined and measured as the combination of the time series of the system parameters, such as the time series measured at two different nodes in a network that represent interlinked microscopic elements (see Fig. 4.1).

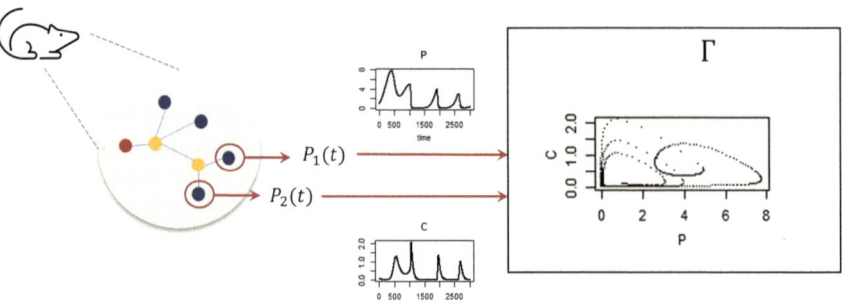

Fig. 4.1 The phase space contains trajectories Γ that are calculated from the time series measured at specific endpoints in complex systems, for example, the concentration of two proteins in a signal transduction network of cells belonging to an organoid of an organism

Note that this approach assumes that the small world is somehow complete, i.e., that interacting elements have stable states that are accessible from an objective point of view and can be arranged within well-defined spaces. With such a concept, mathematical tools, such as group theory, can be applied to these spaces.

We realize that such a mathematical representation departs from ideal interacting elements with well-characterized states σ_i. For example, an electron s_i can have intrinsic angular momentum with two opposite states that can be represented via a binary system $\sigma_i = \begin{cases} + \\ - \end{cases}$.

These interacting elements are coarse-grained representations of objects or agents with complex internal structures and internal reservoirs that are systematically ignored to maintain a low degree of complexity.

This mathematical idealization of the interacting objects is thus a reduction in complexity (small world), which aims to represent other complex systems as a product of the aggregation of such simple interconnected elements (complexity derived from simplicity). In this way, one or more scales are ignored to mathematically represent a scale of interest.

Although the reduction of any interacting elements into simple elements, such as particles (without dimensions and internal characteristics), has been successful in several examples to mathematically describe complex systems, there are still persistent problems regarding such complexity reduction.

The main problem is that despite all the efforts to mathematically represent complex systems with sophisticated models, many systems are still not fully predictable (from markets to biological systems). Is it possible that the method used to dissect the big world into small worlds is limited?

Our goal is to introduce a different and more general type of interacting element that is highly dependent on context. In this approach, we define context as the reference scale in which elements define their interactions with their neighbors. Therefore, it may have been a mistake to deviate from the assumption that complex systems are complete and that they lead to natural laws that are based on absolute coarseness. Instead, we need to accept that complex systems are irreparably incomplete.

4.2 Topological Multiscaling

The purpose of this chapter is to present the problem in an intuitive way so that readers with less technical mathematical background are able to understand the concept of incompleteness in complex systems.

To better understand this problem, we introduce mathematical concepts into what we define as topological multiscaling. The question is as follows: is the topology of a space or a set absolute, or does it depend on the shape of the elements contained in that space? If the topology of a space is not absolute, is it possible for the elements to define the topology of the space?

Mathematical models usually assume that a space is absolute and that the elements contained in that space can be abstracted into ideal objects such as particles. In this context, an element is any object that is contained in a space and can interact with other elements.

An element can be a point (an abstraction of a complex object without an internal structure that interacts with other points through direct contact), a particle (interacts through a field), or an agent (with inherent structures and complex interactions with other agents).

Nevertheless, there is no mathematical principle that prevents the existence of variable forms of both the contained elements and the spaces. Both forms can be defined relative to each other. The object (interacting element) can therefore redefine its "shape" on the basis of a local context and change the reference space. Thus, the reference space takes on a different topology (e.g., a compact space) in response to the original topology. Consequently, it is impossible to define the topology of mathematical spaces completely (see Fig. 4.2).

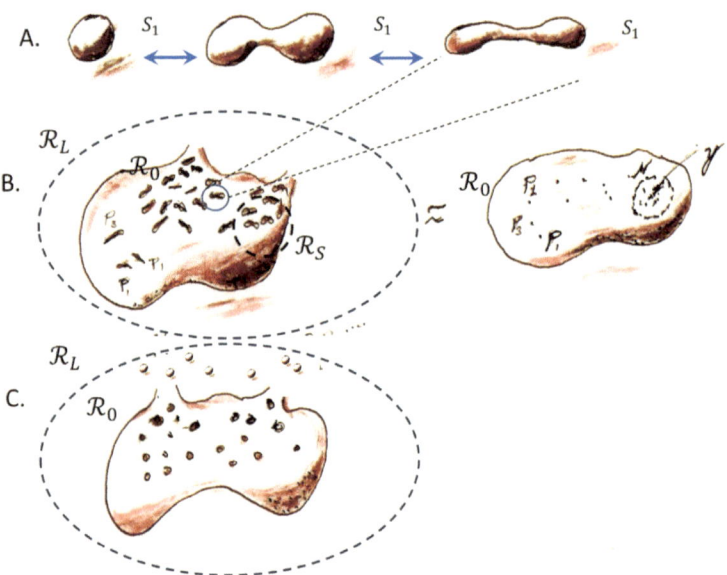

Fig. 4.2 An intuitive definition of topological relativity. Elements enclosed in a space can adopt different shapes (**a**). The elongated (peanut) shape indicates that space $\mathcal{R}_0$ is compact, i.e., it can be covered by subspaces $\mathcal{R}_S$ and is not included in its complementary space $\mathcal{R}_L$ (**b**). In such a case, the contained elements cannot detect open regions in this space. However, a change of the element's shape into a small sphere implies that, suddenly, the open elements of the space are detected, since particles can also be present in the complementary space $\mathcal{R}_L$; this also implies that the space $\mathcal{R}_0$ is no longer compact (**c**). In this example, we consider that the definition of the element's shape is relative to the space $\mathcal{R}_0$ (peanut shaped) or the space $\mathcal{R}_L$ spherical. In such cases, the particles are enclosed in a concatenated space $\mathcal{R}_0 \subseteq \mathcal{R}_L$, which is a form of topological multiscaling

The following mental experiment helps to better understand this concept: consider single elements s_i belonging to the space $\mathcal{R}$. These elements are defined relative to a specific context that we can designated as K.

Thus, we will refer not only to the element but also to how this element is constrained by its context, for instance, the reference to a certain scale; we represent this situation in a symbolic way as $s_i \rfloor_K$.

Now, assume that such states can be related to topological characteristics, such as their shape, depending on a particular context. Therefore, such topological characteristics can change $s_i \rfloor_K \rightarrow s_i \rfloor_{K'}$ depending on different contexts K or K' without violating its fundamental topological invariance.

Different shapes preserve the essential element's topology, such as its genus, which intuitively means counting the number of holes or cavities on a surface (a concept used, for instance, in topological data science for the analysis of persistent homology; see, for example, Emrani et al. [15]).

In Part a, Fig. 4.2, the elements on the right (which look like peanuts) respond to a context with subspaces with genus $G(0)$. For $s_i \rfloor_K$, these elements belong to $\mathcal{R}_0$, which is a compact space (Fig. 4.1b). However, for $s_i \rfloor_K \rightarrow s_i \rfloor_{K'}$, the space $\mathcal{R}$ is no longer compact since the change in the shape of the elements allows their flow from "tunnels" that were invisible to the elements with the initial shape $s_i \rfloor_K$ (Fig. 4.2c).

The simplest and most "natural" topology of the elements in topological space should be a particle without a specific shape. However, the topology of the space in this example is heavily influenced by the shape of the elements it contains.

Therefore, the space cannot be compact (the boundaries are open) depending on how the elements bound in this set "observe" this space.

While mathematical spaces are conventionally defined as a set that contains points that are not dependent on a given context, in this example, we introduce spaces that are not neutral to the context, for example, how an **element** $s_i \rfloor_K$ forms its shape depending on a scale of reference, such as the larger space $\mathcal{R}_L$ in Fig. 4.1. Here, we stress the difference between **particles (small-world concept)** and **interacting elements (large world)** owing to their inherent nonreducible characteristics extending across several scales.

A **scale** is defined regarding a single set $\mathcal{R}$ here, assuming that this set is a subset of a larger set $\mathcal{R}_L$ and that there is an infinite concatenation among the sets and subsets. By this, we are able to contextually define spaces and their topological properties, calling them topological multiscaling.

According to this definition:

- From the perspective of an objective observer, the state σ_i of the object s_i belongs to a space $\mathcal{R}_0$, i.e., $s_i \in \mathcal{R}_0$, regardless of the scale of the system, such that $s_i \notin \mathcal{R}'_0$, $\mathcal{R}_0 \subseteq \mathcal{R}_L$ and $\mathcal{R}'_0 \subseteq \mathcal{R}_L$.
- From the single-state perspective s_i, there is a preferred scale that provides the context to s_i, $s_i \rfloor_K$, with respect to a preferred space $\mathcal{R}$.

On the basis of this definition, states must be defined in the context with the appropriate scale (and not just as a mathematical element in a space) because such a context cannot always be objectively retrieved. Thus, the following axioms can be defined:

A multiscale is defined as a continuous concatenation of different subsets, i.e., (A1)
$$\ldots \subseteq \mathcal{R}_0 \subseteq \mathcal{R}_1 \subseteq \ldots$$

A state σ_i of an object s_i in a multiscale system belongs to a reference space, i.e., (A2)
$s_i \in \mathcal{R}$, and has a specific identity K in respect to a scale Υ, $s_i \rfloor_K$, i.e., $\sigma_i \rfloor_K \in s_K$.

A trivial interacting element has reference to a single space (mathematically $s_i \in \mathcal{R}_0$), whereas nontrivial interdependent elements at multiple scales can both have different reference spaces (mathematically $s_i \rfloor_K \in \mathcal{R}_0$ and $s_i \rfloor_{K'} \in \mathcal{R}_0'$).

This last condition implies that an interacting element cannot be defined in an absolute way in relation to a given space and can possess two or more identities at the same time. Several identities imply that the same element can simultaneously have different kinds of states $\sigma_i \rfloor_K$ and $\sigma_i \rfloor_{K'}$, i.e., $\sigma_i = \{\sigma_i \rfloor_K, \sigma_i \rfloor_K'\}$.

For example, while in physics, it is common to reduce complex objects (molecules, organisms, agents, etc.) into particles with well-defined states, such objects can have intrinsic structures (and therefore reservoirs) that, depending on their context, induce a change in identities, implying elastic and undefined states. As a result, both the context of the viewer and the context of the interacting element affect the system dynamics.

According to the previous definitions, the objects that are defined while taking into account A2 cannot be trivially classified into classical statistical ensembles, such as microcanonical or canonical ensembles, as we have objects whose associations with an ensemble depend on their context.

Consequently, ensembles are only applicable to particle-like objects without an intrinsic structure or with a negligible intrinsic structure (limited multiscale).

To illustrate this case, Fig. 4.2 shows how a single element can have a variety of identities: an element $s_i \rfloor_K$ may have either a complex shape (peanut shaped) or a spherical shape $s_i \rfloor_{K'}$, similar to simple particles, depending on the element's context (see Fig. 4.2).

In the first case, consider the following example: the seeds of some plants, including burdocks, have microscopic hooks that can attach to clothing and animal fur. This allows them to spread geographically. Let us say that we are doing an experiment on a sloped, smooth surface where these seeds roll like a small ball.

In this mental experiment, the context (the smooth surface) defines the identity of the element so that the seed behaves like a small ball. In another case, if the context is a forest, the seed tries to stay at any host (a bird or a fox) that would help passively disperse them (seeds) from one location to another.

There is no way to define the identity of the element in an absolute term, which is therefore a form of incompleteness (see Fig. 4.3).

This change in identity is similar to the elasticity of words. However, language, especially the language of youth, is very dynamic, and its meaning can change

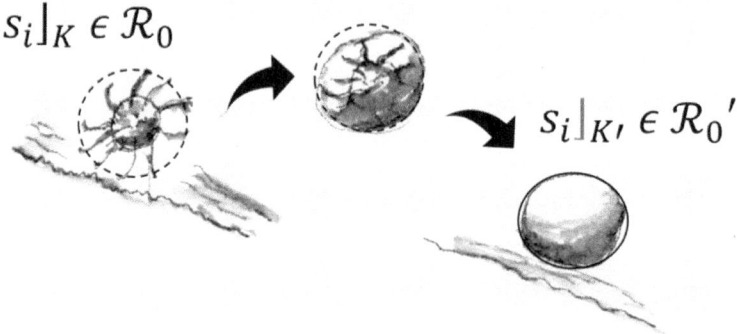

$$s_i \rfloor_K \in \mathcal{R}_0$$

$$s_i \rfloor_{K'} \in \mathcal{R}_0'$$

Fig. 4.3 Coarse-grained representation of complex interacting elements. The elements s_i can assume different entities depending on the context, i.e., if the surface has a structure (like a cloth) or if the surface is smooth. Thus, the element's identity $s_i \rfloor_K$ depends on the context K, which simultaneously influences the states $\sigma_i \rfloor_K$ that can adopt this element

quickly depending on the context and user group. For example, the word "basic" is synonymous with "essential" ($\sigma_1 \rfloor_K$). In youth language (and for some groups), "basic" can mean boring ($\sigma_1 \rfloor_{K'}$). In fact, meaning is constantly changing depending on how it is used and in what context. Therefore, emergent language models based on pattern analysis of a language can hardly reflect static patterns in well-defined contexts [4].

Typically, in a small-world approach, the context is defined as the interaction of the state s_i with other neighboring states s_j, external forces or fields; all these interactions can be studied objectively.

However, the context of this example is not only a product of the interactions among the elements but also how they are inherently perceived.

Thus, according to axiom A2, such a context is related to σ_i and its reference scale, which is not necessarily equivalent to the reference scale of the observer. Assuming that σ_i is not trivially reducible and is embedded into a family of different scales Υ_n, then $\sigma_i \rfloor_K \rightarrow \sigma_i \rfloor_{\langle K \rangle}$, i.e., for complex states embedded in a multiscale, there is a probability that a state is not exactly constrained to a given context K in an absolute way but related as a whole family of contexts $\langle K \rangle$.

4.3 The Fundamental Problem of Complex Systems

In the previous sections, we argued that there is persistent incompleteness and that interacting elements are not necessarily stable, thus making them theoretically indescribable. What are the consequences of such definitions in complex systems and in systems theory? In the next subsections, we attempt to answer this question.

4.3.1 Uncertain Entities of Interacting Elements

In a complex system, the fluctuation/dissipation of the system limits its dynamics and collective interactions. Complex systems are thus located at an interface that includes collective interactions on several scales.

However, deriving objective descriptions of complex systems, which are usually based on physical principles, often ignores the role of context for individual elements and their states (equivalent to choosing an appropriate depth for the interface).

In the previous section, we discussed the relevance of context in multiscale spaces via a simple mathematical concept. In this description, there is constant incompleteness when objects and their corresponding states are set in spaces with multiple scales. According to A1, multiscaling does not necessarily refer to the aggregation of objects but rather to concatenated subspaces.

There are several sets and subsets that serve as references to the element within the context. Accordingly, from A1 and A2, the following theorem can be derived[1] (**persistent incompleteness**):

An interacting element can adopt multiple possible identities depending on multiple (T1)
contexts $\ldots \subseteq \mathcal{R}_0 \subseteq \mathcal{R}_1 \subseteq \ldots \mathcal{R}_1 \subseteq \ldots$ *and can adopt a variety of identities and*
states as

$$\sigma_i = \{\sigma_i\rfloor_K, \sigma_i\rfloor'_K, \ldots\} = \{[(\sigma_i\rfloor_K)_1, (\sigma_i\rfloor_K)_2, \ldots], [(\sigma_i\rfloor'_K)_1, (\sigma_i\rfloor'_K)_2, \ldots], \ldots\}$$

Thus, each state can take different but well-defined deterministic values $(\sigma_i\rfloor_K)_l$ depending on its context. However, for a context-dependent element, it is not possible to derive a single well-defined state.

This contrasts with the implicitly assumed completeness of physical microstates, which can take different values and can be accessed in an objective way, i.e., $\sigma_i = [(\sigma_i)_1, (\sigma_i)_2, \ldots]$ (for example, in time series). This enables its deterministic description as soon as all boundary conditions are mathematically formulated.

Considering that the context cannot be assessed and that the local definition of this context follows a constant pattern (i.e., periodic incompleteness), it should be possible to capture all distributions of the state $\sigma_i\rfloor_K$, i.e.,

$$\sigma_i = \langle \sigma_i\rfloor_K \rangle \tag{4.1}$$

In such a case, the sample $\langle \sigma_i\rfloor_K \rangle$ can be described deterministically. We chose to use a formalism inspired by quantum mechanics, where the single vector $|\sigma_i\rfloor_K\rangle$ represents all the context available to the single object. This could mean that there are different available contexts in quantum mechanics and that the fundamental problem in this field is the impossibility of objectively defining observables[2] [31].

---

[1] This one is similar to the consistency problem in the first Goedel's theorem https://plato.stanford.edu/entries/goedel-incompleteness/.

[2] Quantum mechanical experiments have a limited number of contexts, and therefore a sample can be defined. The context of the problem described in this section is not well defined, so it is not possible to capture all possible states.

4.3.2 Elastic States, Complex Systems and Statistical Mechanics

A small-world approach (context-independent) is very convenient, as interconnected elements can practically be reduced to particles with quantifiable and accurate states. Monomers can be reduced to particles that interact through a potential or particles with intrinsic magnetic momentum, which can be expressed by an array of particles with discrete states in a network or lattice, as an example.

Similar approaches implicitly assume that the interacting elements are immutable and that any internal structure is negligible. We can also develop models that focus on the emergence of macroscopic states via the interaction of a single element via this coarse-grained representation. As a result of this reduction, it is possible to represent different systems (e.g., social, biological, or physical systems) with similar models [8, 18, 21].

Given that interacting elements are not trivially reducible, such interacting objects first require a clear context and a locally preferred scale that cannot be objectively determined.

Concepts of emergence are based on well-defined objects whose interactions lead to macroscopic states. In addition, complex systems are not subject to fundamental laws or principles per se. As a result, the system creates its own conditions according to its collective dynamics. All of these concepts can be related to the concept of elastic states:

The state $\sigma_i|_K$ belonging to the context-dependent interacting elements $s_i|_K$ is   (D1)
defined as elastic state, i.e., is a state that can change its local reference from the
space $\mathcal{R}_0$ to a space $\mathcal{R}_j$ from a family of concatenated spaces $\ldots \subseteq \mathcal{R}_0 \subseteq \mathcal{R}_1 \subseteq \ldots$

It is not possible to trivially reduce or define complex systems on the basis of the absolute definition D1. Rather, these states elastically adjust on the basis of their preferred reference space.

On the one hand, our experience shows us that the world seems more or less complete. However, the persistent incompleteness of interacting elements and states implies that there should be mechanisms capable of compensating for this incompleteness, thus, in turn, making systems complete, at least locally.

The need to define a context to produce consistent particles could imply that interacting elements could eventually "calculate" their context, or similarly, their own decision-making mechanisms, even primitive ones, to solve such persistent incompleteness and create at least a local consistent space (this aspect is discussed in Chap. 6).

We propose here that the balance between persistent incompleteness and local completeness requires the notion of fundamental decision-making, since information is not simply a state but an elementary element required to decide on the referential context of a given object to determine its state. In the case of vesicles, for example, a change in identity in the system (replication inhibition relative to the environment) can stop infinite vesicle replication.

Inherent decision making allows the definition of a preferred identity in respect to a   (D2)
concatenated space $\ldots \subseteq \mathcal{R}_0 \subseteq \mathcal{R}_1 \subseteq \ldots$, *such that* $\{\sigma_i \rfloor_K, \sigma_i \rfloor'_K, \ldots\} \to \sigma_i$, *which*
allows the definition of a preferred space or context $\mathcal{R}_j$ *from a concatenated family*
of spaces/contexts, i.e., $\{\ldots \subseteq \mathcal{R}_0 \subseteq \mathcal{R}_1 \subseteq \ldots\} \to \mathcal{R}_j$.

Only in this case can the system be described deterministically. Therefore, completeness is not given; it is not absolute, inherent or objective but rather defined by the system itself. In this way, we propose that computing is much more fundamental and linked to basic natural processes, which are not limited to artificial systems and cannot be trivially reduced to physical laws [12].

According to Mitchell, *computing is a natural science that may eventually serve as the foundation for biology, just as physics has served chemistry. That is, the science of computing may someday contribute to the conceptual building blocks upon which a more unified understanding of biological phenomena may be based* [26].

This implies that information can play more pivotal roles than storing biochemical information for replication in complex systems. To understand complex systems, not only information (as a form of neg-entropy [28]) but also its content and meaning in combination with computations are essential [12].

The need for decision-making in a complex system requires basic decision-making skills that also require cognitive skills. According to definition D2, decidability is not a consequence but a fundamental property that is needed in complex multiscale systems in itself.

This type of primordial cognitive ability can be associated with what is defined as basal cognition[3] [23], a concept that has been studied in several biological systems showing cognitive capabilities and decision making (such as unicellular fungi finding their way out in a complicated labyrinth) without having sophisticated brain structures.

Thus, basal or primordial cognition seems to be a fundamental property required in simple organisms and relies on fairly simple structures [23], which is a concept that matches what we introduced in D2.

As we have shown in the previous chapter, concepts from statistical physics have proven useful for deriving models and principles for complex systems, including causal properties, which often rely on decomposition into simple elements that can be abstracted into abstract concepts, such as the concept of particles in physics, which are context independent (small-world approach).

On the other hand, context plays a fundamental role (large-world approach), and in the case of persistent incompleteness, fundamental decision-making is required not only as a physical consequence—as an emergent phenomenon—but also as an essential element to solve persistent incompleteness and thus generate local coherent systems, i.e., consistent element definitions with consistent states

---

[3] See also https://www.spektrum.de/inhaltsverzeichnis/intelligent-ohne-gehirn-gehirn-und-geist-9-2024/2199934.

$\{\sigma_i \rfloor_K, \sigma_i \rfloor'_K, \ldots\} \rightarrow \sigma_i$. Only under such circumstances can theories about complex systems be formulated.

4.3.3 Some Implications of Elastic Systems in Complex Systems

What are the consequences in systems theory when interacting elements are elastic? We believe that such a definition could lead to some fundamental changes in the way complex systems are understood.

The laws of thermodynamics, derived from statistical physics and mathematically described via particle samples, apply to a small-world approach since systems can be broken down into elementary interacting elements that can be reduced to particle-like elements, i.e., elements without relevant internal structures are context independent.

From this starting point, the usual method for analyzing complex systems is to study their internal interactions and their consumption of resources such as energy.

Cancer cells differ from other cells not only in that they do not "cooperate" with other cells in the tissue but also because they initiate proliferation and avoid apoptosis (cell death). Above all, their metabolism is essentially based on excessive glucose consumption, which is impaired by the metabolism of other cells in the tissue [17].

Additionally, societies are usually analyzed in terms of their resource consumption: chimpanzees are often considered aggressive ape species with respect to other species, such as bonobos (which have a high level of social cohesion), in part because chimpanzees have to plan and expend considerable energy to access their food sources, which is stressful for this ape community. Bonobos are quite different: you will find them relaxing under the canopy of the rainforest, where there are abundant food sources.[4]

In all these different systems, however, the influence of context cannot be ignored, which means that the notion of physical statistical systems can be a kind of oversimplification for both physical and nonphysical systems, i.e., the persistent tendency to disregard the inherent properties of interacting elements and to define abstract definitions that are as precise as possible. Resource consumption is one of the main limitations of the system.

However, the interpretation of the context by the elements of the system and its influence on resource consumption cannot be ignored indefinitely.

By analyzing sample distributions of simplified elements, we are able to define the entropy required not only in thermodynamics and statistical physics but also in complexity measurements. The recognition of more general and contextual interacting systems now challenges the way entropy is defined: a system must first determine its context (or preferred space) before it can optimize or minimize its entropy.

---

[4] See for instance this information: https://www.swr.de/swr2/wissen/bonobos-missverstandene-menschenaffen-swr2-wissen-aula-2021-02-28-100.html.

Therefore, according to axioms A1 and A2, it is not possible to define an ensemble (in statistical physics) in an appropriate way in other more general complex systems.

In light of all these aspects, it is possible to ask the following question: are there fundamental laws governing living systems? This possibility has been explored in Chap. 3, and it has been speculated that there may be fundamental laws, not only for living systems [16] but also for any complex system as a whole.

However, after examining the possibility that interacting elements are elastic, we might conclude that perhaps the answer is that there are no universal and fundamental laws for complex systems.

Looking at the principles of elastically interacting elements, there are probably no universal laws but laws that are the result of a consensus between several interconnected elements.

Therefore, complex systems must be recognized as interacting entities that behave like observers calculating their context rather than simply as objects.

This notion implies the following:

(S1) In multiscale living systems, energy consumption and dissipation define the   (D3)
structures associated with system's context,
and
(S2) According to D2, the preferred system's identity with respect to a preferred
context constrains the structures necessary for consuming and dissipating energy
within the system.

In other words, causal properties depend on the identity of the system and the preferred context and, in many cases, cannot be objectively evaluated.

This has profound implications for applications of statistical physics in complex systems.

For example, complex networks can be context dependent, which implies that complex networks' edge distributions are not necessarily guided by universal laws [6]. Thus, trajectories may be defined not only by energy constraints but also by the systems themselves (according to D2).

4.3.4 Definition of the Distortion Module in Complex Systems

What are the consequences of the theory presented in the previous section? The term D2, which is based on the mathematical definition of elastic states and the implicit 'nondecidability' of complex systems, implies that large-world systems (context aware) require not one but two types of constraints: a physical constraint, which refers to a control function related to resource consumption, and a computational constraint, which refers to decision-making in relation to a given context (which complements the idea that computation is not just a property of artificial systems, as discussed in the previous paragraphs [12]).

This second constraint implicitly requires an inherent system's decision-making capabilities (or basal cognitive capabilities).

Of course, the ability to decide sets limits on the growth of complexity [19]. One way to focus on this problem is to consider that interacting elements not only follow a trajectory bounded by the energy landscape but are also able to compute this trajectory (according to definition D2).

In this framework, we define a trajectory via the fundamental definition of dynamical systems, where $\Gamma^\lambda|_{\{max,min\}U^\lambda}$ is the trajectory of the microstates constrained to the optimization (maximization or minimization) of a control function $\mathcal{H}^\lambda$. This can be the state of a particle in a dynamic system, a concentration or any appropriate observable, such that Γ^λ represents the state of the whole system.

The structure $\hat{M}$, $\mathcal{P}(\hat{M})$ determines how the micro states change over time, i.e., $\mathcal{P} : \sigma_{i,t} \to \sigma_{i,t+1}$, where $\hat{M}$ is a structure in which information is stored and where $\mathcal{P}(\hat{M})$ is the set of rules used to estimate the transition from one input state to the next.

Therefore, $\mathcal{P}$, defined as a computation, codifies the rules required to estimate the next state $\sigma_{i,t}$ (output) using the information stored in the network structure $\hat{M}$. The computation $\mathcal{P}(\hat{M})$ is apparently isomorphic to the system dynamical description. However, from a strictly mathematical point of view, both are completely different mathematical objects [7]. When this computation represents the environment ε, then (see Example 1 in Table 4.1)

$$\mathcal{P}(\hat{M}) = \varepsilon \tag{4.2}$$

Following the definitions introduced by Hernández-Orozco et al. [19], the organism's information is codified in a structure $\hat{M}$, which contains several microstates ordered such that they can respond to external inputs. This structure can be represented by a network such that $\hat{M} = g_{ij}\sigma_i \cdot \sigma_j$, where g_{ij} is the connectivity factor of the information of microstates σ_i and σ_j.

"In weakly convergent systems, the function $\mathcal{P}(\hat{M})$ represents an organism, a theory or any other computable system that uses the structure $\hat{M}$ to predict the trajectory in the environment ε" [19]. If this representation cannot be completed in a certain time period, the system is nonadapted, or the theory is useless.

Owing to the overlap between computational theory and dynamic systems, the computation of the environment ε refers to the computation of the trajectory Γ that represents the environment λ, i.e., $\varepsilon^\lambda = \Gamma^\lambda$ (for this notation, see Wang et al. [34]). This process can be seen as the search for food in a grid or the expression of proteins in a genetic network expressing a particular phenotype.

If the environment is dynamic, i.e., $\varepsilon(\delta_t)$ for a sequence of times $\delta_1 \cdots \delta_t \cdots$, then a computation $\mathcal{P}_t$, such that $\mathcal{P}_t(\hat{M}_{\delta_t}) = \varepsilon(\delta_t)$ exists, or in other words, the existence of a trajectory $\Gamma(\delta_t)$. If a series of representations of the environment $p : t \mapsto p_t$ is computable, the function $\delta : t \mapsto \delta_t$ is computable, and the descriptive complexity of the system is bounded [19].

On the other hand, if the sequence of times $\delta : t \mapsto \delta_t$ is noncomputable, then the system is also noncomputable, i.e., the environment is noncomputable.

Table 4.1 Representation of the control function for systems with low and high degrees of autonomy [14]

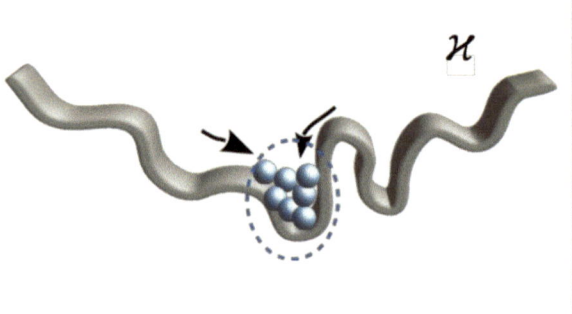	Example 1: Homeostasis represents the optimization of landscape $\mathcal{H}$ in relation to a control function (for instance the fitness of organisms within a population). Both the individual elements (small spheres) and the entire organism (sphere population enclosed in a dotted circle) explore their environment ε using their trajectories Γ, and are constrained to the optimization of the control function $\{max, min\}\mathcal{H}$
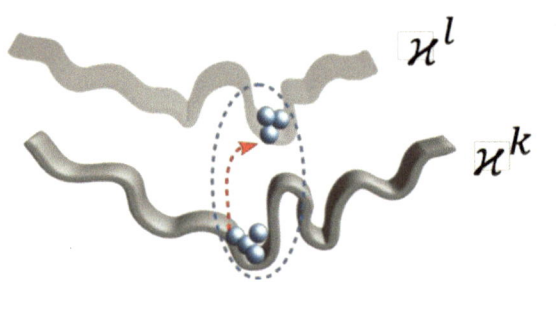	Example 2: The control function $\mathcal{H}^k$ is locally optimized, but the entire population is eventually constrained to a different control function $\mathcal{H}^l$. The gap between these trajectories depends on the accommodation and assimilation of the organism (the population) to this environment, or equivalently on the use of compensation mechanisms so that biological functions are determined

Therefore, there is no way to decide if $\Gamma(\delta_t)$, with its sequence of times representing the environment, implying its complexity growth. In the above definition, open-ended evolution (OEE) represents nature's creative productivity [19, 27].

For noncooperative organisms, $\mathcal{P}_t$ cannot find computable conditions to represent their dynamic environment, increasing the system complexity. These organisms are sensitive to $\varepsilon(\delta_t)$ but are blinded to further fluctuations in the environment. However, organisms can cooperate with other organisms and the environment, and the increase in complexity can be life-threatening (e.g., nonapoptotic cells in cancerous tissues).

For this reason, we argue that systems not only possess OEE but also look for computability conditions.

If more than one (dynamic) environment is calculated, the computability of the system can be imposed. Under this condition, the system explores a family of trajectories $\Gamma^\lambda(\delta_t)$ searching for computable conditions in the environment and generating a biological function in this manner. In multiscale systems (cells, organelles, organs, etc.), this is a plausible scenario, so that

$$\mathcal{P}\_i\big(M^{\lambda}(\delta_t)\big) = \Gamma^{\lambda}(\delta_t)\big|_{\{max,min\}\mathcal{H}^{\lambda}}, \tag{4.3}$$

where $\Gamma^{\lambda}(\delta_t)\big|_{\{max,min\}\mathcal{H}^{\lambda}}$ is the trajectory of the microstates constrained to the optimization (maximization or minimization) of the control function $\mathcal{H}^{\lambda}$ related to the environment λ, which simultaneously belongs to a family of environments $\lambda = 1 \cdots \Lambda$ (see Table 4.1).

This condition implies a distance between two trajectories belonging to two different environments k and l (k and $l \in \lambda = 1 \cdots \Lambda$), which is the absolute value of the mean distance between the points of two trajectories:

$$\Delta^{kl}\Gamma = \left| \Gamma^{k}(\delta_t)\big|_{\{max,min\}\mathcal{H}^{k}} - \Gamma^{l}(\delta_t)\big|_{\{max,min\}\mathcal{H}^{l}} \right|. \tag{4.4}$$

In this expression, there is an implicit difference between different landscapes of control functions due to changes in the environment (see Table 4.1).

These types of systems possess "ears and eyes" that are able to sense dynamic changes in the environment and impose decidability on a process since switching from one relative environment to another stops the computation while reducing the inherent increase in complexity. In contrast, conventional dynamical systems focus solely on calculating a single trajectory.

This also implies the existence of a family of structures, such that $\hat{M}^{\lambda} = g^{\lambda}_{ij}\sigma_i \cdot \sigma_j$. If two trajectories exist, then there is a probability of establishing a distance between different structures such that for two environments k and l, with $\Delta^{kl}\hat{M} = H\left(g^{k}_{ij}, g^{l}_{ij}\right)$ being the hamming distance between these structures.

To describe this phenomenon, the theory of elasticity in solid and liquid matter can be useful when computational structures are deformed to explore new trajectories and make computational systems decidable. We suspect that we can measure the deformation of these structures/mechanisms in the network.

Like the modulus of elasticity in solid mechanics [22, 30], the modulus of elasticity of mechanisms (MEM) is measured as the distance between two network structures with respect to two environments $\Delta^{kl}\hat{M}$ relative to the mean distance between two trajectories $\Delta^{kl}\Gamma$ (Eq. 4.3) and is defined as[5]

$$D^{kl} = \frac{\Delta^{kl}\hat{M}}{\Delta^{kl}\Gamma}. \tag{4.5}$$

In a family of environments, and like the tensor notation of the modulus of elasticity in solid mechanics, expression (4.5) becomes a tensor-like structure:

---

[5] This relation is valid since we relate two distances. No additional mathematical structures are here involved.

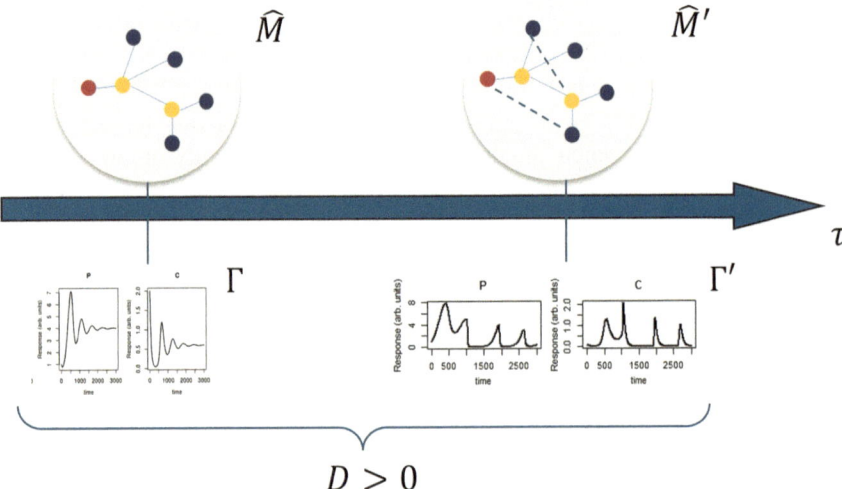

Fig. 4.4 Schematic representation of the effects of changes in the inherent system structure and system dynamics, represented in trajectory Γ. In such cases, time series are extracted within two different time periods

$$
D^{kl} = \begin{pmatrix} \frac{\Delta^{11}\hat{M}}{\Delta^{11}\Gamma} & \frac{\Delta^{\lambda_1\lambda_2}\hat{M}}{\Delta^{\lambda_1\lambda_1}\Gamma} & \cdots & \frac{\Delta^{1l}\hat{M}}{\Delta^{1l}\Gamma} \\ \frac{\Delta^{21}\hat{M}}{\Delta^{21}\Gamma} & \frac{\Delta^{22}\hat{M}}{\Delta^{22}\Gamma} & \cdots & \frac{\Delta^{2l}\hat{M}}{\Delta^{2l}\Gamma} \\ \vdots & \vdots & \cdots & \frac{\Delta^{kl}\hat{M}}{\Delta^{kl}\Gamma} \end{pmatrix}. \tag{4.6}
$$

If $D^{kl} = 0$, there are no changes in the structure, which means that the system remains either noncomputable while retaining the OEE or that the system is complete where simple trajectories can be calculated.

This last case implies that the system can be mathematically described (for instance, via differential equations).[6] On the other hand, if $D^{kl} \gg 0$, there are changes in the structure $\hat{M}$ to make the system decidable (an intuitive graphical representation of this concept is presented in Fig. 4.4).

When an organism cannot calculate its environment or when its trajectory is undecidable, the relevant interacting elements try to modify its internal structures to meet the needs of the organism. In this case, the system is distorted.

In the next few sections, we want to introduce some toy models to better understand the mathematical implications of these ideas.

---

[6] Both conditions are dangerous, since dynamic systems could be used to describe OEE systems. This can be wrong if the gap between the trajectories describing the environment is not recognized.

4.4   Toy Model: Context-Dependent Ising Model

What does the definition of a distortion module have to do with the calculation? An effective way to illustrate this concept is to use a toy model. Mathematical ideal models consist of interacting objects with very simple states arranged in a periodic lattice.

An example of a simple model is the Ising model, which is presented in Chap. 3. By using a periodic grid in this model, we can avoid effects associated with more complex geometries (such as phase transitions or complex order states).

As we also observed in Chap. 3, similar models are also used to represent societies. For example, individuals who are influenced by their neighbors and have to choose between two options, i.e., possess a binary state, σ_i (which can be either $+1$ or -1).

The probability that an individual whose states are influenced by its neighbors will enter one state or another depends on a probability function limited by the sum of the neighboring interactions represented by the Hamiltonian function $\mathcal{H} = \sum_{i,j} g_{ij}\sigma_i\sigma_j$. The resulting dynamics can then be represented with a Master equation, as introduced in Chap. 3 (Eq. 4.5).

Thus, the function $\mathcal{H}$ is the causal property of the system. However, here, there is something very special: the coupling function g_{ij} encodes the system's geometry.[7] As in physics or in any representation of biological processes, coupling refers to the process in which one element or individual interacts with another, in this case its neighbors.

Such interactions can either be constant or evolve depending on the system dynamics, i.e., the coupling is no longer constant and changes over time $g_{ij} = g_{ij}(t)$ (see Fig. 4.5).

In the present example, however, we introduce a different type of context: a lattice with imperfect coordination [29]. The grid is similar to several works by MC Escher with impossible geometries. An "imperfect coordination" metric represents interrupted connectivity between observers at different scales provided by a common connectivity operator.

For example, the context of an interacting system can be defined by recognizing that the interacting elements are also observers, thus forming a common background reference space (i.e., the background reference space is shared by all observers; see Part a in Fig. 4.5).

In the present case, ζ_l defines the degrees of freedom for the coordination between two observers seeking their common background, such that $g_{ij} = \zeta_i \hat{g} \zeta_j$, where $\hat{g} = \begin{pmatrix} +1 & r(+) \\ r(-) & -1 \end{pmatrix}$ is a matrix, with $r(+)$ and $r(-)$ being randomly defined between 0 and $+1$ as well as 0 and -1, respectively, across the lattice.

Thus, the element i seeks its own context with respect to the given context $\zeta_i \hat{g}$ and has a transition probability that depends on the average neighbors, such that the control function is defined as $\mathcal{H} = \sum_{\langle i,j \rangle} \left[\zeta_i \hat{g} \zeta_j \right] \times \sigma_i\sigma_j$. While in a simple Ising

---

[7] And plays a similar role as a metric in general relativity.

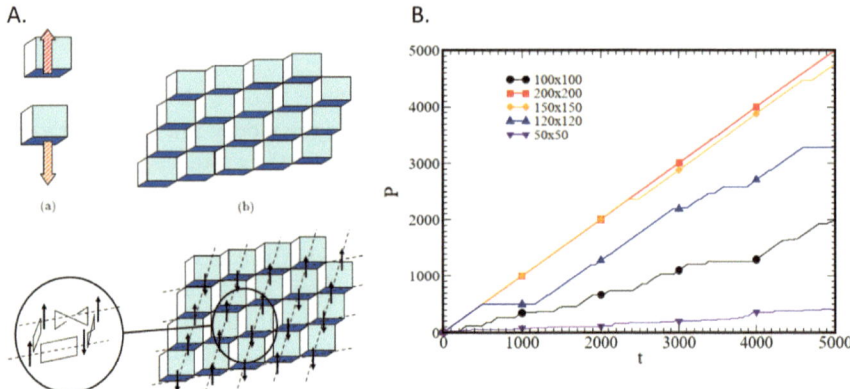

Fig. 4.5 This toy model is a representation of a lattice that is defined according to the context of the expected element (**b**). This lattice has been inspired by well-known graphical illusions (for example, those sketched by M. C. Esher). In this lattice, the coupling between the elements has an orientation (±) depending on the context of how the lattice is locally interpreted (**a**). When an Ising model is defined on the basis of this lattice, the elements can therefore locally interpret their lattice according to their preferred local context. Reprinted from Ramirez Barrios and Díaz Ochoa [29], with permission from World Scientific

model, g_{ij} is a constant in the entire lattice, in this experiment, g_{ij} can be a constant or a variable constant (observe the similarity between this equation and Eq. (3.8) for the application of game theory in solid-state systems).

A relevant finding of this experiment is that the coupling or metric derived from this definition is not simply a function of time but is determined by the interacting elements themselves, as well as by a more general context that cannot be fully objectively evaluated.

The results obtained from imperfect lattice couplings suggest that there is a kind of punctual equilibrium in the topology of the lattice, depending on the lattice size and cannot be easily sampled: while small lattices exhibit relative stability or a tendency toward an interrupted equilibrium (which has been visualized by calculating the cumulative average value of the couplings as a function of time, Part b of Fig. 4.5), larger lattices tend to become unstable and cannot be defined as stable couplings.

This indicates both the variability of a system and a limitation in recognizing probable causal patterns, as shown in Fig. 4.5 [29].

This computer experiment highlights the importance of the perspective of both the observer and the elements that interact (irregardless of how primitive these states may be).

Using a line of reasoning used in quantum mechanics, the mental state of the observer influences the observed system in terms of the way the system can be forced to be observed[8] [31]. Moreover, being able to recognize the implicit mental states of interacting elements implies that they can decide their interactions in their own respective contexts.

4.5 Example: Vesicle Replication

In this first example, we consider the formation of vesicles from an aqueous system containing the surfactant S [14]. In this case, a highly lipophilic precursor of S, known as $S–S$, is hydrolyzed at the vesicle boundary.

As a vesicle grows, it eventually divides into two or more thermodynamically stable vesicles.

The more vesicles that are formed, the more $S–S$ are bound, creating more vesicles, making the process autocatalytic. Considering that the entire hydrolysis and growth process takes place within and around the boundary, the vesicle can be seen as a simple, self-replicating autopoietic system [3].

A simple vesicle undergoes internal reactions when a membrane (S) is formed from molecule B through a process with a generation velocity v. As a result, S decays with the velocity of degradation v_{deg}. Let us also assume that the precursor metabolite A enters the environment and that B breaks down into C, which is eventually expelled.

This cycle can be infinite if the process is homeostatic and has access to sufficient energy sources. At this point, all these processes are mechanistic and can be described via simple ordinary differential equations (see Fig. 4.6).

On the basis of Maturana and Varela's work, the interaction between an autopoietic unit and its environment can change if basic cognition concepts are considered [24, 25]. Indeed, whenever something occurs, there is adaptation:

(a) To an environmental cause (in the example of the vesicles, an outer molecule X–Y);

(b) A resulting effect from the unit (here, the inception of the molecule X and the release of the metabolite Y).

Moreover, the effect has an adaptive virtue, inhibiting vesicle reproduction and causing degradation to occur [3]. As a result, this cycle continues indefinitely.

The fact that the bound molecules X–Y can trigger or inhibit internal processes (within the unit) implies the introduction of an external decision about the next state of the vesicle in relation to the population of other vesicles and the environment.

In this example, biological processes cannot make decisions "on their own" on the basis of only mechanistic (and physical) ideas at the micro level. We suspect that this problem occurs in many systems in biology and other sciences.

---

[8] Which is a form of solipsism.

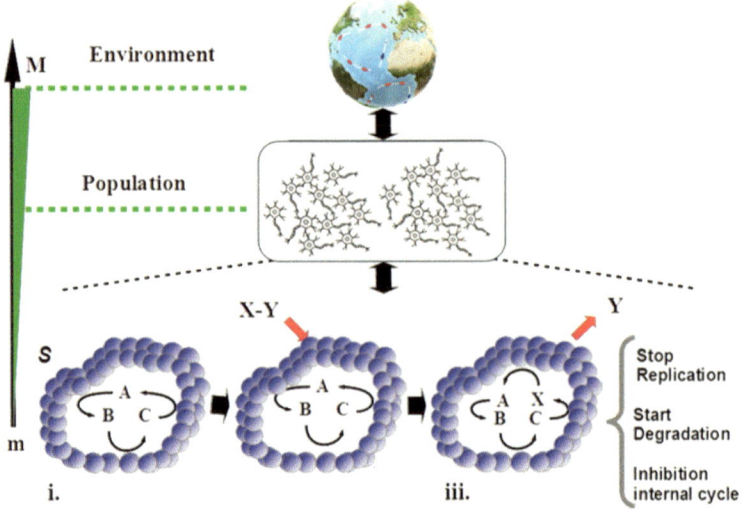

Fig. 4.6 Example of a vesicle with a membrane formed by a cycle generated with molecules A, B and C (**i**) that allows vesicle replication. After interaction with the nearest population as well as with the environment, a pair of molecules X–Y is absorbed (**ii**); thereafter, the molecule X is linked into cycle (**iii**). This interaction can trigger important "decisions" in the vesicle concerning its maintenance or degradation or the inhibition of the internal cycle. Reprinted from Diaz Ochoa [14], with permission from Springer Science + Business Media

The undecidability of "at infinity" in 2D lattices of atoms means that even if the spectral gap is known for a given finite lattice, it can change abruptly—from gapless to gap-like spectral or vice versa—as the size increases, even by just a single atom.

Since it is "probably impossible" to predict when—or if—this is the case, general conclusions can be drawn from experiments or simulations [9, 11]. Notably, we observe here that concepts from computer theory have been extended to physics.

Our hypothesis is that the internal properties of vesicles act like changes in the internal states of atoms, so populations of vesicles cannot stop on their own. Accordingly, biological processes require assimilation and accommodation.

In other words, the dynamics of the system (in this case, replication) are set to come to a standstill (i.e., stop the replication process) by a self-contained definition of its elements and conditions.

This leads to confusion about the identity of the system. As a result, we propose that microbiological systems are inherently incomplete and that assimilation and accommodation allow them to become complete and decidable.

For example, if the population of vesicles has enough energy, they can and will eventually multiply (like an immortal cancer cell in a tumor), which is no different from the infinite iterations of a Turing machine. If this unit does not belong to an organism, this process is determined according to the functionality of the population as a whole.

4.6 Example: Adapting the Predator–Prey System as a Toy Model

The last section illustrates the possibility of understanding systems as inherently incomplete. Of course, we should consider the practical consequences of this theory. To this end, we explore these concepts with a toy model of a biological system and evaluate the implications for mathematical modeling.

The system in this example consists of a population of organisms with chemotactic reactions that are eaten by a second population. In this system, the context is created by the population of predators, allowing an organism to change its identity by switching between two regulatory mechanisms to change the dynamics and adapt to the context [14].

Because intrinsic networks are not consistent, there are several possible answers that correspond to different network architectures. To illustrate this, we use the example of chemotaxis, where it is difficult to decide between two candidate networks associated with two different responses to stimuli [10].

We argue that the system is noncomputable in the sense that it cannot decide on its own which course should be calculated.

A signal response that is perfectly adapted to its environment seems to be quite limited in terms of the number of possible ways to achieve this. It has been proposed that signaling networks are divided into two classes depending on their architecture or topology: those with a negative (integral) feedback system and those with two parallel paths that first diverge and eventually converge, affecting outputs in opposite directions. The latter type of network has been referred to as the "incoherent feed-forward loop" [1].

In this instance, adaptation to temporally changing inputs can be the key to this response (Fig. 4.7).

A model for a population has been developed on the basis of this architecture for the organism's response. Figure 4.7 illustrates the input $z(t)$, intracellular concentrations for activation and inhibition loops ($x(t)$ and $y(t)$), and response $r(t)$ of the organisms, as well as the development of the population of consumers $C(t)$ and predators $P(t)$, assuming that the population behaves according to Lotka–Volterra theory.

Here, the equations for the chemotactic response are adapted for the responses of a social amoeba [33]. The results for each response are presented under the assumption that stimuli linearly increase with time to avoid artifacts that other stimuli can introduce [10, 14].

The cells maximize their response as the rate of change increases while recognizing that they are approaching the source of the chemoattractant. However, some organisms may require adaptation to external reactions.

For example, when they approach the source, they minimize their response by requiring the NFB architecture instead of the IFFL architecture. As a result, it can be difficult to determine or define the underlying network. In addition, we observe

System 1:
Switch between system 2 and system 3

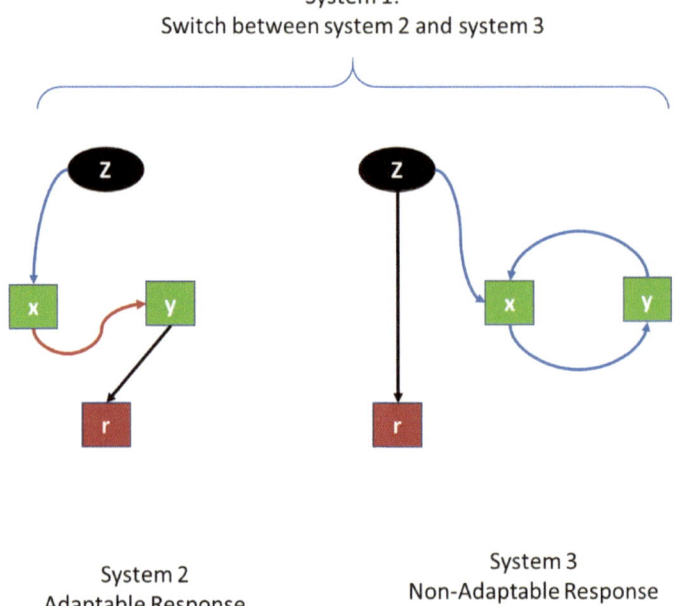

System 2
Adaptable Response

System 3
Non-Adaptable Response

Fig. 4.7 Example of chemotaxis where a population must decide between two different kinds of responses: one for an iFFL architecture g_{ij}^1 (left) and an NFB g_{ij}^2 architecture (right), depending on the interaction of the entire population with the environment (predator's population)

an incomplete system, i.e., two different competing models that ultimately rely on common molecular interactions (Fig. 4.8).

We argue that both reactions are characteristic of cells that can adapt to their environment. Therefore, there is not only the reaction of an organism (which is the conventional way models are defined). The dynamic creates a new context that leads to a change in identity for the components of the system.

Furthermore, this implies that the a priori chemotactic organisms are unknown, which means that the system could, in principle, be incomplete.

Consequently, the cells as a whole decide which response is most suitable, depending on the behavior of the population of consumers and predators (Fig. 4.8). In this way, we establish a direct interaction between individual organisms and the entire population.

With this concept, we construct a toy model via the following steps [14]:

- Assume that consumers ($C(t)$, population of predators) are close to a chemical stimulus frequented by a population of bacteria that is predated ($P(t)^9$). In our example, the stimuli grow in proportion to time.

---

[9] Here the function $P(t)$ refers to the predator's population, which is different to $\mathcal{P}(\hat{M})$, which is defined as the computation using the information encoded in $\hat{M}$.

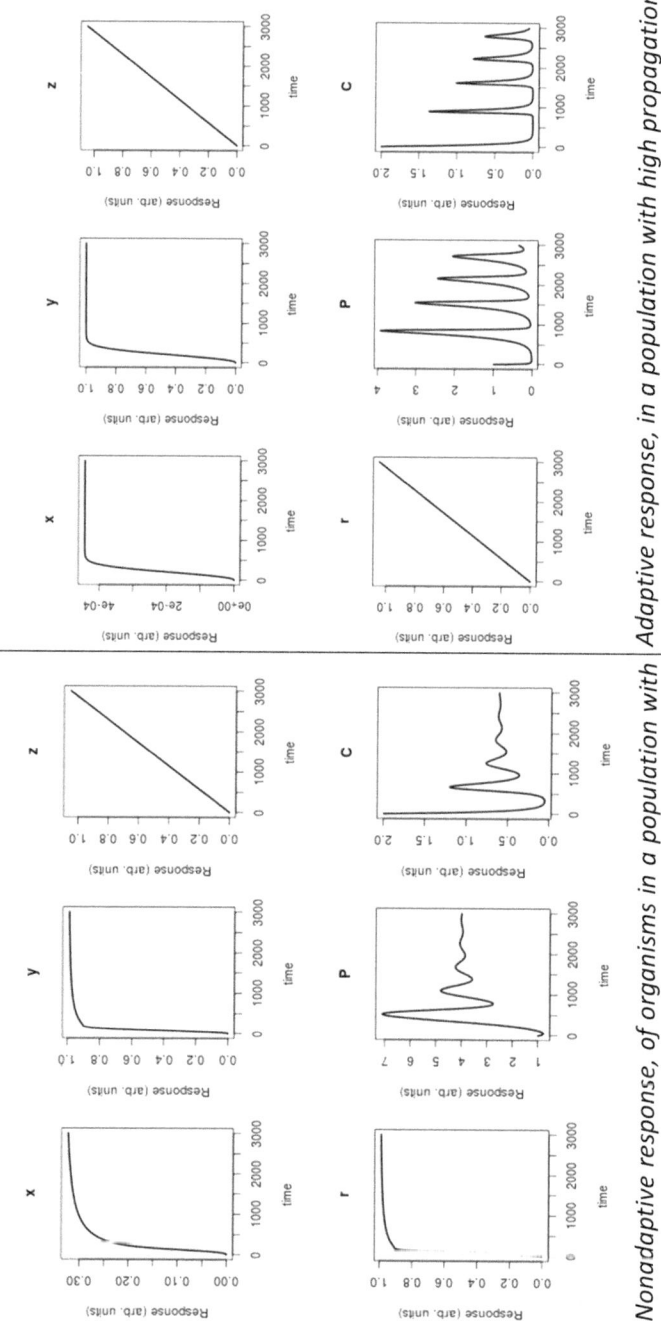

Nonadaptive response, of organisms in a population with low propagation velocity $v(t)$. The equations describing the organism's response, based on a NFB architecture

Adaptive response, in a population with high propagation velocity $v(t)$. The equations describing the organism's response, based on an iFFL architecture

Fig. 4.8 Population dynamics of organisms with adaptive and nonadaptive responses. Reprinted from Diaz Ochoa [14] with permission from Springer Science + Business Media

- The bacteria react variably: they switch from an adaptive to a nonadaptive reaction. Bacteria alone cannot "decide", which is a better response; this decision is made as an organism but takes into account the entire population and its environment. The response will depend on the entire predator population.
- If $C(t)$ (population of predators) is relatively low, the preferred response remains constant; otherwise, if the population of predators increases above a critical number, the preferred response will change from a nonadaptive to an adaptive response.
- The type of response influences the velocity toward the stimuli: a nonadaptive response is related to a high speed of propagation in the direction of the stimuli. Otherwise, an adaptive response is associated with a slow rate of propagation (see Experiments on Changes in Speeds for Bacterial Aggregation in Chemotaxis).

The interacting elements change their identity and switch from one reaction to another (similar to the example of polymer adsorption). Through this "trick", prey organisms can influence the number of predators and thus also influence the number of populations that occur [2].

In this example, there seems to be coevolution. However, we find that the identity of the organisms is not fully decided (between an adaptive and a nonadaptive response) and that the underlying network is therefore incomplete, as it has two potential background models competing with each other.

The apparent coevolution is the decision to adopt one or more responses depending on the pressure on the entire population of consumers and predated organisms. In this process, there is constant distortion until the organism meets a "decision" and selects only one response. According to Eq. (4.5), the first $\tau_a = 1000$ time steps are as follows:

$$
\begin{aligned}
H\left(g_{ij}^1; g_{ij}^2\right) &= 2 > 0, \\
|r(t) - r_{S=1}(t)| &> 0, \\
D^{12}|_{t<\tau_a} = D^{21}|_{t<\tau_a} &> 0.
\end{aligned}
\tag{4.7}
$$

The system drifts into critical behavior, where the cells take turns selecting two of the answers. The change in response depending on the population of the predator implies a bias:

$$
D^{12} = D^{21} > 0
$$

The additional oscillation of this distortion implies self-organization in the population dynamics and adaptation of the response. Additionally, in this model, we assume that each kind of response is complete, i.e., $D^{11}|_{t<\tau_a} = D^{22}|_{t<\tau_a} = 0$.[10]

---

[10] Otherwise, we must assume for instance more than one adaptive response associated to different models.

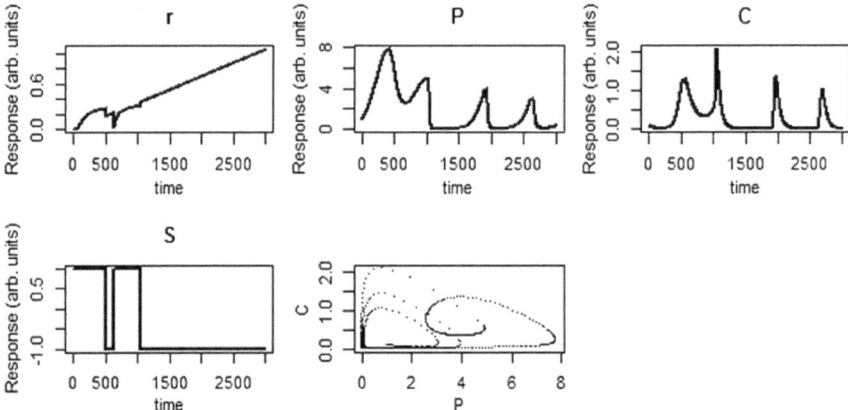

Fig. 4.9 Figure illustrates the predated population $P(t)$ (in our example bacteria) and the consumers $C(t)$. In this case, S detects whether a nonadaptive ($S = 1$) or adaptive response ($S = -1$) has been selected. Reprinted from Diaz Ochoa [14], with permission from Springer Science + Business Media

Above a critical value of the stimuli, the cell finally selects one response from the two potential responses (after 1000 time steps), implying that $D_{21}|_{t>\tau_a} = D_{12}|_{t>\tau_a} = D_{11}|_{t>\tau_a} = D_{22}|_{t>\tau_a} = 0$. This transition is visualized in the phase diagram in Fig. 4.9.

In this example, there is uncertainty about how to choose the most appropriate response to the environment. Microstates do not simply react to the environment but actively select a reaction (in our example, from two possible reactions) and modify their behavior at higher levels. In this way, higher levels close the loop and create completeness, i.e., multiscaling is responsible for completeness.

As a result, we have a theoretical example of how organisms change their identity depending on the context in which they develop. Therefore, the organism is able to stop or restart a computation (a response or function) in relation to its macrostate reaction.

This also implies that the organism can, in principle, maintain different redundant structures that can iterate either the same response or different responses, a decision that depends on multiscaling where the organism is embedded [32].

4.7 Last Remarks

According to both the theory and the examples presented in this chapter, we can conclude that there is no absolute representation of the identities of interacting elements since these identities depend on the context in which they exist.

The states of an element cannot always be described objectively and are elastic in nature. Therefore, extracting causal paths from complex systems is much more difficult than extracting noise (even assuming that regularities in the data allow automatic extraction of causal pathways [5]).

The most serious limitation in systems theory is its fundamental mathematical incompleteness: systems theory does not define the context directly, as there is not only an indirect interaction across scales (as has been shown with the concept of emergence in the previous section) but also a persistent incompleteness that requires direct interaction/decision-making across scales.

In short, information plays a crucial role in defining the context at hand.

As we have shown in the various examples, this could mean that in multiple systems, it is not possible to access an absolute representation of causal properties represented by a function (or field) $\mathcal{H}$, which can be highly distorted, considering that its function depends on a distorted structure $\hat{M}$ (Eq. 4.5). It seems that optimization and efficiency are also limited in understanding biological systems if such an argument is true.

For instance, the classical understanding of Darwinist evolution is based on the concept of a fitness function $\mathcal{H}$ that leads to the survival of the fittest individuals: a living being can only survive if it adapts as perfectly as possible to its environment, which leads to a slavish dependence on an objective external world defined by function $\mathcal{H}$. As a result, the environment and identity of the organisms are precisely defined.

However, this cannot be assumed owing to the context, i.e., the reference scale. Rather, the concept of identity in relation to a context requires an inherent (subjective) perspective that cannot be easily reduced [24].

Changing identities and elastic states therefore pose a significant challenge for deductive and inductive modeling. This includes basic models trained on large datasets [4]. This limits the ability to extract universal principles or properties from complex (biological) systems.

In the next chapter, we address a more precise quantification of this limitation.
Stuttgart—Schwarzenberghütte

References

1. Alon U (2006) An introduction to systems biology: design principles of biological circuits, 1st edn. In: CRC mathematical & computational biology. Chapman & Hall, Boca Raton, FL
2. Arditi R, Ginzburg LR (1989) Coupling in predator-prey dynamics: ratio-dependence. J Theor Biol 139:311–326. https://doi.org/10.1016/S0022-5193(89)80211-5
3. Bitbol M, Luisi PL (2004) Autopoiesis with or without cognition: defining life at its edge. J R Soc Interface 1:99–107.
4. Bommasani R, Hudson DA, Adeli E, Altman R, Arora S, von Arx S, Bernstein MS, Bohg J, Bosselut A, Brunskill E, Brynjolfsson E, Buch S, Card D, Castellon R, Chatterji N, Chen A, Creel K, Davis JQ, Demszky D, Donahue C, Doumbouya M, Durmus E, Ermon S, Etchemendy J, Ethayarajh K, Fei-Fei L, Finn C, Gale T, Gillespie L, Goel K, Goodman N, Grossman S,

Guha N, Hashimoto T, Henderson P, Hewitt J, Ho DE, Hong J, Hsu K, Huang J, Icard T, Jain S, Jurafsky D, Kalluri P, Karamcheti S, Keeling G, Khani F, Khattab O, Koh PW, Krass M, Krishna R, Kuditipudi R, Kumar A, Ladhak F, Lee M, Lee T, Leskovec J, Levent I, Li XL, Li X, Ma T, Malik A, Manning CD, Mirchandani S, Mitchell E, Munyikwa Z, Nair S, Narayan A, Narayanan D, Newman B, Nie A, Niebles JC, Nilforoshan H, Nyarko J, Ogut G, Orr L, Papadimitriou I, Park JS, Piech C, Portelance E, Potts C, Raghunathan A, Reich R, Ren H, Rong F, Roohani Y, Ruiz C, Ryan J, Ré C, Sadigh D, Sagawa S, Santhanam K, Shih A, Srinivasan K, Tamkin A, Taori R, Thomas AW, Tramèr F, Wang RE, Wang W, Wu B, Wu J, Wu Y, Xie SM, Yasunaga M, You J, Zaharia M, Zhang M, Zhang T, Zhang X, Zhang Y, Zheng L, Zhou K, Liang P (2022) On the opportunities and risks of foundation models. https://doi.org/10.48550/arXiv.2108.07258

5. Briggs WM (2024) The unfulfilled quest for discovering cause from probability. In: Ngoc Thach N, Kreinovich V, Ha DT, Trung ND (eds) Optimal transport statistics for economics and related topics. Springer Nature Switzerland, Cham, pp 38–44. https://doi.org/10.1007/978-3-031-35763-3_2

6. Broido AD, Clauset A (2019) Scale-free networks are rare. Nat Commun 10:1017. https://doi.org/10.1038/s41467-019-08746-5

7. Buescu J, Graça DS, Zhong N (2011) Computability and dynamical systems. In: Dynamics, games and science I. Springer, Berlin, Heidelberg, pp 169–181

8. Capra F, Luisi PL (2014) The systems view of life: a unifying vision, 1st edn. Cambridge University Press

9. Castelvecchi D (2015) Paradox at the heart of mathematics makes physics problem unanswerable. Nature 528:207

10. Chang H, Levchenko A (2013) Adaptive molecular networks controlling chemotactic migration: dynamic inputs and selection of the network architecture. Philos Trans R Soc Lond B Biol Sci 368:20130117. https://doi.org/10.1098/rstb.2013.0117

11. Cubitt TS, Perez-Garcia D, Wolf MM (2015) Undecidability of the spectral gap. Nature 528:207–211

12. Denning PJ (2007) Computing is a natural science. Commun ACM 50:13–18. https://doi.org/10.1145/1272516.1272529

13. Diaz Ochoa JG (2014) Relative constraints and evolution. Int J Mod Phys C (IJMPC) 25:1–10

14. Diaz Ochoa JG (2018) Elastic multi-scale mechanisms: computation and biological evolution. J Mol Evol 86:47–57. https://doi.org/10.1007/s00239-017-9823-7

15. Emrani S, Gentimis T, Krim H (2014) Persistent homology of delay embeddings and its application to wheeze detection

16. England JL (2015) Dissipative adaptation in driven self-assembly. Nat Nano 10:919–923. https://doi.org/10.1038/nnano.2015.250

17. Fadaka A, Ajiboye B, Ojo O, Adewale O, Olayide I, Emuowhochere R (2017) Biology of glucose metabolization in cancer cells. J Oncol Sci 3:45–51. https://doi.org/10.1016/j.jons.2017.06.002

18. Galam S (2012) Sociophysics: an overview of emblematic founding models. In: Galam S (ed) Sociophysics: a physicist's modeling of psycho-political phenomena. Springer US, Boston, MA, pp 93–100. https://doi.org/10.1007/978-1-4614 2032-3_5

19. Hernández-Orozco S, Hernández-Quiroz F, Zenil H (2016) Undecidability and irreducibility conditions for open-ended evolution and emergence. arXiv E-prints 1606, arXiv:1606.01810

20. Hoefer C (2023) Causal determinism. In: Zalta EN, Nodelman U (eds) The Stanford encyclopedia of philosophy. Metaphysics Research Lab, Stanford University

21. Jensen HJ (2023) Complexity science: the study of emergence, New. Cambridge University Press, Cambridge, United Kingdom, New York, NY

22. Landau LD (2004) Theory of elasticity. Butterworth-Heinemann Ltd.

23. Lyon P, Keijzer F, Arendt D, Levin M (2021) Reframing cognition: getting down to biological basics. Philos Trans R Soc B Biol Sci 376:20190750. https://doi.org/10.1098/rstb.2019.0750

24. Maturana HR, Varela FJ (1980) Autopoiesis and cognition: the realization of the living, 1980th edn. Springer, Dordrecht, Holland, Boston

25. Maturana HR, Varela FJ (2004) el Arbol del conocimiento: las bases biológicas del entendimiento humano (Santiago de Chile; Buenos Aires: Lumen)
26. Mitchell M (2010) Biological computation. Comput Sci Fac Publ. Present
27. Packard N, Bedau MA, Channon A, Ikegami T, Rasmussen S, Stanley KO, Taylor T (2019) An overview of open-ended evolution: editorial introduction to the open-ended evolution II special issue. Artif Life 25:93–103. https://doi.org/10.1162/artl_a_00291
28. Pagel L (2023) Entropy and information. In: Pagel L (ed) Information is energy: definition of a physically based concept of information. Springer Fachmedien, Wiesbaden, pp 83–115. https://doi.org/10.1007/978-3-658-40862-6_4
29. Ramirez Barrios E, Díaz Ochoa JG (2009) Imperfect coordination in optimization. Int J Mod Phys C 20:527–538. https://doi.org/10.1142/S0129183109013868
30. Rathgeber S (2002) Practical rheology. In: 33. IFF-Ferienkurs 2002, Institut Für Festkörper-forschung: soft matter, complex materials on mesoscopic scales. Schriften des Forschungszen-trum Jülich, Jülich, p C9.2
31. Saunders S (2005) What is probability? In: Elitzur AC, Dolev S, Kolenda N (eds) Quo Vadis quantum mechanics? The frontiers collection. Springer, Berlin, Heidelberg, pp 209–238. https://doi.org/10.1007/3-540-26669-0_12
32. Tononi G, Sporns O, Edelman GM (1999) Measures of degeneracy and redundancy in biological networks. Proc Natl Acad Sci 96:3257–3262. https://doi.org/10.1073/pnas.96.6.3257
33. Wang CJ, Bergmann A, Lin B, Kim K, Levchenko A (2012) Diverse sensitivity thresholds in dynamic signaling responses by social amoebae. Sci Signal 5:ra17. https://doi.org/10.1126/scisignal.2002449
34. Wang Y, Xu M, Wang Z, Tao M, Zhu J, Wang L, Li R, Berceli SA, Wu R (2012) How to cluster gene expression dynamics in response to environmental signals. Brief Bioinform 13:162–174. https://doi.org/10.1093/bib/bbr032

Chapter 5
System Observability and Φ_S Complexity

Keywords Causality · Elastic states · Topological persistence · Integrated information · Causal inference

Mathematical modeling assumes that the object's identity is well known and can be accessed at any time. However, in the previous chapter, we formalized the concept of elastic states. Elastic states arise from interacting elements with identities that depend on their specific context. According to such a concept, causal effects and causal properties can also be elastic.

If a system is observable and controllable, it is possible to predict its response to an input. Otherwise, elastic states imply that it is not always possible to observe the system with precision, setting a practical limit for both deductive and inductive mathematical modeling.

Throughout this section, we introduce a complexity measure according to which low complexity indicates the observability of the system. In essence, this measure of complexity, called Φ_S complexity, provides information about both the inherent autonomy and variability of a system by estimating a persistent entropy in the measured data. With this method, one can determine the degree of causality between systems and visualize the boundary between objective understanding and observability.

In addition, this method can be used to automatically extract features from time series data so that machine learning models can be trained.

This chapter provides an introduction to the mathematical foundations required for this complexity measurement, as well as the background required for its interpretation.

5.1 Causal Events, Causal Properties, Observability and Control

If how the system reacts to disturbances is known, it is then possible to predict, modify and control its reaction and evolution.

In Fig. 5.1, correlations and patterns are determined by analyzing the measured signals, e.g., to identify possible relationships between measurement parameters. According to Reichembach's principle, two correlated variables must have a common cause: either one is a cause for the other or there is a third variable that is a common cause for both [26]. According to Chap. 3, the third variable can be either another event (or a set of intrinsic events) or a causal property.

This leads to a deterministic understanding of the system, where analysis of the correlation between two observables is needed. In this framework, observing how a system behaves when critical parameters (such as temperature) are changed or investigating how disturbances or external changes (input signals) change the state of the system is relevant.

As discussed in Chap. 3, causal inference (either as causal properties or as causal events) can be performed by evaluating how past measured/observed events affect present and future events. Unlike the mechanistic approach presented in the previous chapter, in which elastic states influence their connectivity and thus the adjacency matrix in complex networks (leading to the definition of a distortion factor), in this chapter, we focus more on a black-box approach, in which observed data are used to derive potential causal relationships, with correlations disappearing when probabilities attributed to the common cause are bound [26] (see Fig. 5.1).

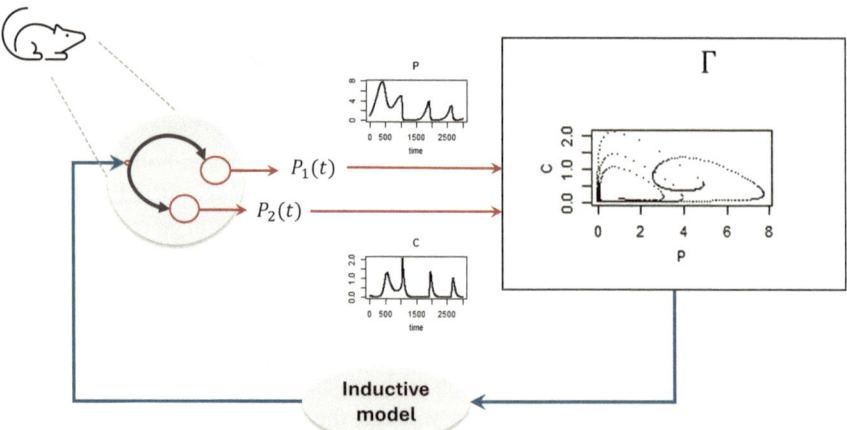

Fig. 5.1 Reconstruction of a black-box model of a system (for example, a cell) on the basis of data analysis aimed at establishing correlations that allow causal inference between different observables and thus the identification of a deterministic system (contrast this figure with Fig. 4.1 in Chap. 4)

Causality refers to the relationship between observed data associated with implicit mechanisms (common cause), which can be used to characterize deterministic properties in observed systems. The causality extracted from data can thus be mathematically described as

$$X(t) = \sum_{\tau=1}^{L} A_\tau (X(t - \tau)|X(t))X(t - \tau) + \varepsilon(t), \qquad (5.1)$$

where in general, $A_\tau (X(t - \tau)|X(t))$ represents a matrix or a mathematical structure connecting observations from the past ($X(t - \tau)$) with observations in the future ($X(t)$).

As a result, methods such as Kalman filters or Granger causality are based upon this basic concept (Runge et al. 2023); additionally, causality tests can be applied to time series to assess causality [22].

Furthermore, we observed in Chap. 4 that these causal relationships, even in directed graphs, have a significant effect on how the system responds to changes in parameters or input signals. In essence, this is the collective effect of the overall system on the individual states of its interconnected components.

In general, understanding such responses, as well as the structure of the system, opens the door to effective control of the system, which has profound implications at all levels, from systems biology or systems medicine (e.g., predicting the most efficient treatment for cancer) to sociology and economics (e.g., predicting consumer behavior). The ability to control complex systems makes mathematical modeling extremely valuable.

Many modern technologies cannot function without mathematical modeling, and there is a constant perception that systems can be observed, predicted, and controlled with increasing accuracy.[1]

However, there are still significant challenges in interpreting system observability and controllability. First, systems are variable, and second, their states may be elastic in relation to their context.

The examples presented in Chap. 4 illustrate how elastic states affect output signals. However, the problem is not the generation of models but rather the theoretical background necessary to understand the data. Signals are subject to persistent defects due to elastic states and their effect on the system response.

In general, there is evidence that persistent incompleteness across several scales limits our ability to deduce causal pathways and our ability to know and control complex systems [6, 30]. Thus, as suggested in Fig. 5.1, we aim to recognize defects in recorded data that differ from mere stochastic effects and noise and, by doing so, assess whether a system is deterministic (no elastic states) and thus modellable or if defects in the data suggest intrinsic elastic states in the system (large world) that could limit the system's observability.

---

[1] There are still serious ecological, social and economic problems pointing to the fact that systems are not controllable as intended. The evidence points to the fact that science and technology are suitable at disrupting more than controlling the world.

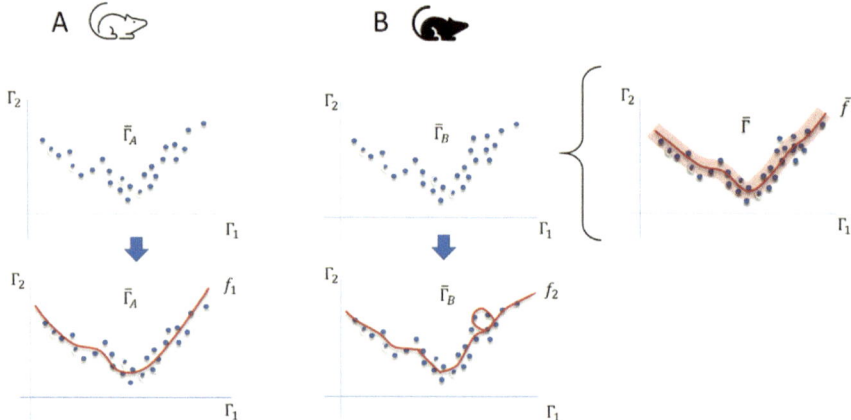

Fig. 5.2 Comparison of the sampling trajectories in phase spaces generated by environmental changes or modifications to the internal regulatory mechanisms of two different organisms, **a** (white mouse) and **b** (black mouse). Among these are the size of a population, electrocardiograms, and concentrations of substances. In **b**, inherent modifications in the organism lead to modifications in the trajectory $\Gamma_1(t)$ that are ignored when sampling is performed. Reprinted from Diaz Ochoa [7]

In the following sections, we introduce novel methods for recognizing elastic states in empirical data.[2]

5.2 Elastic States and Persistent Bias

In Chap. 4, we realized that elastic states entail changes in the trajectories of the system, as elastic states are expected to cause subtle changes, which impairs the ability to extract mathematical models.

Our hypothesis is that mathematical models can be extracted by analyzing the trajectory in phase space, as shown in Figs. 5.1 and 5.2, where a trajectory is sampled in a phase space (each point is a discrete measurement of the trajectory sampled in phase space; individual variations can imply variations in the generated trajectories, as is shown in Fig. 5.2).

A system that has similar mechanisms should show similar responses to external perturbations under similar conditions. In this way, an average value from the phase space can be sampled and used as a basis for training a model.

The presence of similar causal properties implies the existence of mechanistic models with stable physical constraints (for example, graphs–complex network

---

[2] The main content of this chapter has been extracted from the article *"Observability of Complex Systems by Means of Relative Distances Between Homological Groups"* published in Frontiers (physics) [7].

models). On the basis of this assumption, the function $\overline{f}$ is both descriptive, representing the distribution of data points associated with an average path, and predictive, which assists in estimating future responses (the case is intuitively presented in Fig. 5.2, part a). Intuitively, the function $\overline{f}$ is essentially an extracted model.

In the next step, we need to estimate the effect of statistical errors on extracting the model function $\overline{f}$. Statistical errors can be minimized by sampling more data points to minimize the error of the extracted function or model.

Big data methods are also based on this concept to detect regularities in sample datasets. A de facto assumption of big data modeling is that there are stable patterns that can be generalized from any dataset. A large dataset can also be used to reduce bias, which is the statistical error caused by the modeler's assumption, as well as the parameters defined to fit a model to reference data.

However, subtle differences between datasets can go far beyond statistical deviations or outliers in average data samples. As in the example of the two organisms A and B in 2. Such deviations may indicate another physical limitation resulting from a change in the environment of the organism or its internal regulatory mechanisms.

By using a separate analysis, we are able to detect subtle changes in the trajectory, which means that two different models are required for two completely different trajectories, $\overline{\Gamma_A}$ and $\overline{\Gamma_B}$.

For this reason, a method is needed that goes beyond purely statistical fluctuations to extract relative variations caused by changes in physical boundary conditions.

Therefore, we use variance bias to evaluate differences between trajectories with similar responses, leading to the concept of persistent bias, which in turn is related to this sustained variability of physical constraints. In the next section, we describe the mathematical foundations of these concepts in more detail.

5.3   Mathematical Definitions of Persistent Bias and Persistent Entropy in Data

In this section, we present readers with a formal definition of persistent entropy that uses the basic definition of the measured error. At the end of this section, we learn that a measurement error is not solely due to the bias of an observer.

Essentially, there is an inherent and persistent bias in the observed data due to constant internal variations within the system.

According to bias-variance decomposition, the error of a model $\hat{f}$, $Error(\hat{f})$, is composed of three terms: a bias that depends on the researcher's definitions, a variance term, and an inevitable and irreducible error term given by the following equation:

$$Error(\hat{f}) = E\left[\left(\overline{\Gamma} - \hat{f}\right)^2\right] = Bias(\hat{f}^2) + var(\hat{f}) + \sigma^2 \qquad (5.2)$$

where $E[X]$ is the expected value of X, which is the sum of all the possible outcomes of X multiplied by the corresponding probabilities; here, $Bias(\hat{f}^2)$ is the bias of the model $\hat{f}$ (see also Hastie et al. [18]). As explained in the previous section, this bias is the result of incorrect assumptions in the parameters used in the learning algorithm.

However, the dynamics of individual systems, including the autonomous behavior of the system, can lead to persistent bias in the data structure. As shown in Chap. 4, the key argument is that this bias is not introduced randomly by externally defined model parameters but is introduced intrinsically by the system itself (assuming that the system also behaves like an observer):

Theorem   (T1)
In a system k with autonomous intrinsic behavior there is an intrinsic bias,
$Bias\left[\overline{\Gamma}^{kl}\right]$, *defined relative to other similar systems l.*

Therefore, we propose that the measured error is biased by both the experimenter and the system itself.

To demonstrate theorem T1, we consider the differentiation of a system k with respect to other similar systems l. Consider, for example, how internal regulatory processes in an organism k are defined and how they differ with respect to other organisms l. Thus, the variability of the estimated error is defined as

$$\Delta^{kl} Error(\hat{f}) = \Delta^{kl} E\left[\left(\overline{\Gamma} - \hat{f}\right)^2\right] = E\left[\left(\overline{\Gamma}^k - \hat{f}\right)^2\right] - E\left[\left(\overline{\Gamma}^l - \hat{f}\right)^2\right]$$
$$= E\left[\overline{\Gamma}^{k^2} - 2\hat{f}\overline{\Gamma}^k + \hat{f}^2\right] - E\left[\overline{\Gamma}^{l^2} - 2 \cdot \hat{f} \cdot \overline{\Gamma}^l + \hat{f}^2\right]$$
$$= E\left[\overline{\Gamma}^{k^2} - \overline{\Gamma}^{l^2}\right] - 2E\left[\hat{f} \cdot \overline{\Gamma}^k - \hat{f} \cdot \overline{\Gamma}^l\right]$$
$$= Bias\left[\left(\overline{\Gamma}^{kl}\right)^2\right] - 2 \cdot \hat{f} \cdot Bias\left[\overline{\Gamma}^{kl}\right], \tag{5.3}$$

In this expression, the function $Bias\left[\overline{\Gamma}^{kl}\right]$ is defined as the collection of all the expected values of the relative deviations of both trajectories, which mathematically means that the bias is the expected value of the difference of both trajectories, i.e., $Bias\left[\overline{\Gamma}^{kl}\right] = E\left[\overline{\Gamma}^k - \overline{\Gamma}^l\right]$. Thus, considering that $\overline{\Gamma}^k$ and $\overline{\Gamma}^l$ are sets of discrete points, with $\left(\overline{\Gamma}^k - \overline{\Gamma}^l\right) = \{\gamma_1, \gamma_2, \ldots, \gamma_n\}$ and $\left(\overline{\Gamma}^k - \overline{\Gamma}^l\right)^2 = \{\gamma_1', \gamma_2', \ldots, \gamma_m'\}$ also defined as a set of discrete points, then

$$Bias\left[\overline{\Gamma}^{kl}\right] = E\left[\overline{\Gamma}^k - \overline{\Gamma}^l\right] = \sum_{i=1}^{n} P(\gamma_i), \tag{5.4}$$

$$Bias\left[\left(\overline{\Gamma}^{kl}\right)^2\right] = E\left[\left(\overline{\Gamma}^k - \overline{\Gamma}^l\right)^2\right] = \sum_{i=1}^{n} P(\gamma_i'); \tag{5.5}$$

where $P(X)$ is the probability of occurrence of γ_i and γ_i', i.e., the probability of the occurrence of differences between sets k and l of discrete points. This definition of bias is thus equivalent to perturbing the error with respect to the measurement and reference trajectories.

In terms of observability, it refers to the ability to extract the internal states of a system to obtain a coherent and complete description of that system. In this scenario, the extracted model must be stable, and its error must be small enough to be ignored, i.e.,

$$\Delta^{kl} Error\left(\hat{f}\right) = 0. \tag{5.6}$$

In such cases, $\hat{f}$ can describe these internal states and can eventually fulfill the theorem of observability (see, e.g., [21]). This implies:

Theorem (T2)
Systems that can be observed, called observable systems, have no intrinsic bias.

Otherwise, when there is a relative deviation of the observed error of the function $\hat{f}$, $\Delta^{kl} Error\left(\hat{f}\right) > 0$, then there is a probability of detecting persistent defects in the data. Such persistent defects, which originate from autonomous changes in elastic states, imply that there is a probability that significant relative differences arise in data points, i.e.,

$$P(\gamma_i) > 0 \text{ and } P(\gamma_i') > 0. \tag{5.7}$$

In particular, such probabilities allow us to define the intrinsic and persistent entropy of a system, that is, a type of entropy that cannot be trivially removed from the observed system. On the basis of our analysis, we argue that entropy arises from the internal states of the system and can be characterized as follows:

$$H\left[\overline{\Gamma}^{kl}\right] = \sum_{i=1}^{n} P(\gamma_i) \cdot \log(P(\gamma_i)) > 0, \tag{5.8}$$

$$H\left[\left(\overline{\Gamma}^{kl}\right)^2\right] = \sum_{i=1}^{n} P(\gamma_i') \cdot \log\left(P(\gamma_i')\right) > 0. \tag{5.9}$$

A persistent bias, therefore, cannot be simply regarded as a statistical error resulting from the observer's or sampler's actions. Rather, it is the result of persistence entropy resulting from changes in the internal state of a system.

As a result, systems that are observable display a low level of persistent entropy, which can be achieved only if sufficient information is available for observation. This is a logical consequence of the fundamental conditions that define observability.

The next task is to find suitable methods to measure this entropy. We believe that recent advances in persistent topology can be helpful in providing such a measurement.

5.4   Topological Persistence

Since the observed data have patterns with a structure, it can be assumed that such patterns are the result of stable internal states that lead to this behavior. Persistent distortions in the observed systems are expected to lead to defects in the collected data and thus in the recorded patterns. Geometric analysis of the data is useful for formalizing the terms "pattern" and "pattern break" and providing quantitative results.

To achieve this, we use topological data analysis (TDA). Often, datasets are embedded in a higher-dimensional space, and topological information is not well researched. TDA is used to determine the shape of the distributor in which the data are embedded [17].

The following sections provide the reader with an intuitive interpretation of the theory. As part of TDA theory, we also provide a formal definition of persistent homology. The mathematical definitions are then used in a more formal definition of persistent bias in complex systems.

5.4.1   The Intuitive Approach

The likelihood that trajectories measured from similar systems will have significant differences, as described in Eq. (5.6), implies that it should be possible to extract features from these data that could help detect such differences.

Notably, stochastic effects are expected, and such trajectories contain a certain amount of noise that distinguishes them slightly. Therefore, we need a basic notion and method capable of distinguishing between features that characterize the patterns and structures of the different datasets collected and their inherent noise.

Here, we pay special attention to the concept of characteristic structures in measured data rather than patterns, as we refer to how each collected data point is glued together with other data points to form a trajectory (which can be conserved as a whole object).

Similarly, effects induced by autonomic states can be measured as changes in the main structure. This type of measurement is similar to looking at cracks in the structure of a bridge.

Consequently, the objective is to compute the intrinsic bias and intrinsic persistent entropy on the basis of a sample of different relative trajectories $\overline{\Gamma}^{kl}$ and $\left(\overline{\Gamma}^{kl}\right)^2$ [2]. To perform a model or regression, we extract the topological structure of the collected data, ideally by combining different trajectories in a phase space.

Figure 5.3 illustrates this procedure intuitively, sampling trajectories in phase space. The definition of the homology group depends on changes in the trajectories (this case is similar to the example shown in Fig. 5.2).

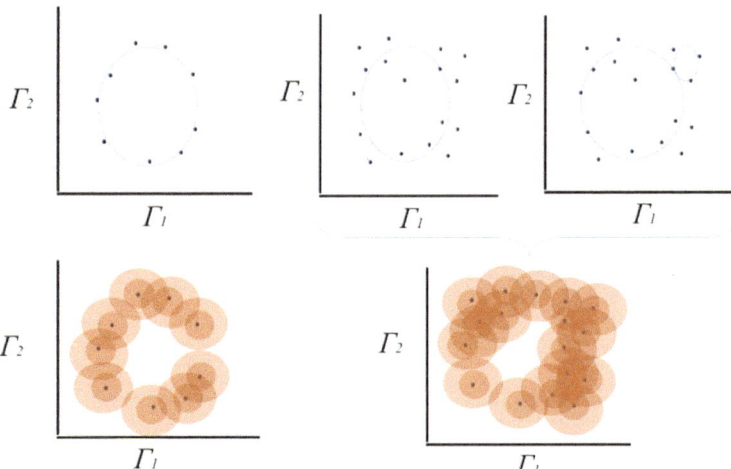

Fig. 5.3 This example illustrates how a proximity parameter can be used to estimate topological persistence. The combination of a set of points can be converted into an Čech complex [16]. The example on the left side shows connected points and a surface with homologies 0 and 1. Using the same procedure, a second structure with homology 1 can be discovered on the right. Reprinted from Diaz Ochoa [7]

As shown in Fig. 5.4, point clouds are generated from a sample of the system trajectory to estimate this structure. Point clouds, mathematically defined as $\overline{\Gamma}^k = \left(\Gamma_1^k(t), \Gamma_2^k(t), \Gamma_3^k(t), \ldots\right)$, consist of data points sampled from the primary form $\left\{\overline{\Gamma}^k\right\}$, i.e., from all the data points obtained from an observed trajectory.[3] By combining the time series of the trajectories, it is possible to recover the dynamics of the system [31]. In addition, by analyzing persistent homology, we can identify harmonic structures in the data represented in point clouds that are related to these dynamics [13].

As a tool in algebraic topology, persistent homology is particularly useful when the scale is not known a priori. This concept consists of describing the topological-geometric structure of the system trajectories as point clouds with finite filtration in a space [12]; this means an index family of subjects with a given algebraic structure rather than smooth manifolds using real values [11].

Hence, it can be viewed as a generalization of hierarchical grouping of topological characteristics of the highest order, which leads to an invariant represented by barcodes [12, 13, 15]. The method is extensively used for data analysis, for example, to extract features from data for machine learning [27] or to analyze biomedical signals [25].

We sample a collection of points in a metric space into a global object defined as the vertices of a combinatorial graph whose edges are determined by their proximity [15].

---

[3] Which are different from the elements enclosed in a space, as described in Chap. 4.

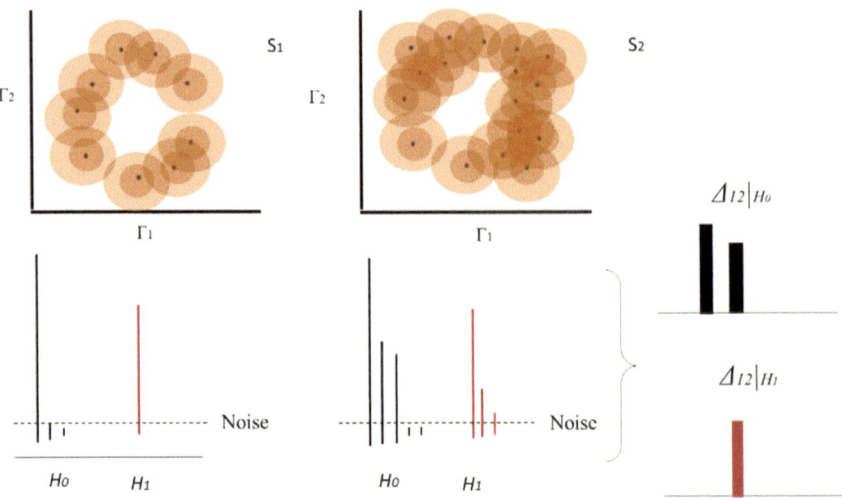

Fig. 5.4 Two different datasets are illustrated in Fig. 5.1 as homology groups and bar diagrams. Each concentric circle represents a distance function. For S1, the persistence diagram provides a homology group H0, which is a group with genus = 1. By increasing the number of points on S2, it is possible to define a homology group H1, which is a surface with connected points. A schematic representation of the difference between the persistent and nonpersistent diagrams is shown on the left. Reprinted from Diaz Ochoa [7]

Although the graph captures the connectivity of the data, simplexes can also be constructed, i.e., clouds of connected spheres defined at each point in $\overline{\Gamma}^k$. The goal is to determine the homology of filtration, i.e., the number of holes exiting the connected data structure. The data structure consists of the point cloud elements that are glued together.

To intuitively understand what this means, we present a quite intuitive illustration of the methodology in Fig. 5.4, where a system S_1 receives a filtration (which is the history of a growing complex, in this case, the "spheres" defined around each data point) that mainly produces connectivity between the points (homology group H_0), whereas S_2 produces not only connectivity but also a structure with a hole, i.e., a structure with a homology group H_1.

Alternately, we may also categorize this method as a type of clustering method that identifies elements that are both close together and exhibit a particular topology.

5.4.2 Mathematical Background

To formalize the previous concepts, it is necessary to define quantifiable topological features, called persistence bars, which measure the persistence of the data structure over time.

For readers who are less interested in the mathematical background, it is enough to recall the intuitive interpretation presented in the previous chapters and move on to the next section on the use of persistent topology to detect inherent biases in measured data.

We first assume that trajectory Γ^λ has a topology that reflects the periodic behavior of a signal with Euler characteristics, which means that Γ^λ has a function g with a compact subset of $\mathbb{R}^D$ and that $d_{\overline{\Gamma}^k} : \mathbb{R}^D \to \mathbb{R}$ is the distance function of $\overline{\Gamma}^k$.

Furthermore, we consider $L = \left\{ \delta : d_{\overline{\Gamma}^k}(\delta) \leq \varepsilon \right\}$ as a set of persistent bars δ_ε that estimates the topological feature length. For example, for a first-order homology group H_1, i.e., a loop in the data cloud, $\delta_\varepsilon|_{H_1}$ in the persistence bar is a measure of how a data point is properly clustered in the group by measuring the distance of the data to the group with respect to the distance parameter ε.

In this context, the barcode is an analog of the Betti number, which is a characterization of the topological properties of a surface. Importantly, the Kth Betti number of a complex acts as a rough numerical measure of the homology group H_k.

The key topological features of H_k include zero (connected points) and first-order topology (loops) (see Fig. 5.3). In the current implementation of a persistent bar, a silhouette plot is used to validate clusters of data points belonging to specific topological features, such as connected components, tunnels and voids, i.e., considering the distance of one point to a cluster C_i^m representing a topological characteristic m; then, the persistence bar can be defined as a feature depending on

$$a_m(i) = \sum_{j \in C_i^m} d(i,j), \tag{5.10}$$

where $d(i,j)$ is the distance from one point i to another point j and

$$b_m(i) = \min_{m \notin i} \sum_{j \in C^n} d(i,j), \tag{5.11}$$

the distance of one point i to a neighbor cluster C^n, respectively [4, 5, 7]. Thus, the persistence bar is defined as

$$\delta_m(i) = b_m(i) - a_m(i), \tag{5.12}$$

which is the distance of the data point to the cluster such that when $\delta_m \to 0$, the point i is near two clusters, whereas $\delta_m > 0$ implies that the data are appropriately clustered and that the data point belongs to a persistent topological characteristic within a homology group H_m [14].

For $\lambda = 1, 2, \ldots \Lambda$, if there is a sample of trajectories generating a point cloud $\{w^\lambda\}$, then $L\{w^\lambda\} = \left\{ \delta_1^\lambda, \delta_2^\lambda, \ldots, \delta_m^\lambda \right\}$ are different topologically persistent characteristics of the trajectory $\overline{\Gamma}^\lambda$ (see Fig. 5.3).

As a result of this approach, a finite set of equivalence classes can be identified, with the property that no trajectory belonging to one equivalence class can be continuously deformed into a trajectory belonging to another equivalence class. This implies that $L\{w^k\}$ and $L\{w^l\}$ are two equivalence classes containing all the relevant information of the inherent structure of the data points of two trajectories $\overline{\Gamma}^k$ and $\overline{\Gamma}^l$.

5.5   Estimation of Inherent Bias and System Variability

In this chapter, persistent entropy and the measurement of topological features are interwoven. As a result, Eqs. (5.8) and (5.9) describe persistent entropy in terms of the topological properties of the observed trajectory. By using such a definition, we can measure the inherent persistent bias in the observed data.

5.5.1   Inherent Bias and Persistent Entropy

With both intuitive and formal definitions, we can now apply persistent homology to detect topological signatures in collected data from trajectory data by simply defining the difference between the equivalent classes of different trajectory data as follows:

$$\left(L\{w\left(\overline{\Gamma}^k\right)\} - L\{w\left(\overline{\Gamma}^l\right)\} \right) = \sum_m \Delta_{kl} | H_m$$

$$= \sum_m \left(\{\delta_1^k, \delta_2^k, \ldots, \delta_m^k\} - \{\delta_1^l, \delta_2^l, \ldots, \delta_m^l\} \right)$$

$$= \mathcal{M}\left(\overline{\Gamma}^{kl}\right) \tag{5.13}$$

where $\delta_{i,m}^k$ is the persistence bar for the corresponding topological feature m, or homology group H_m, of trajectory Γ^k and where $\Delta_{kl} | H_m$ is the total difference between the persistent bars δ_m^k and δ_m^l associated with H_m.

Like the distortion matrix introduced in Chap. 4 (for inherent changing structures in complex systems), we refer to $\mathcal{M}\left(\overline{\Gamma}^{kl}\right)$ as a distortion matrix of observed trajectories (which should not be confused with the system's structure $\hat{M}$).

Notably, the calculation of the equivalence classes is robust to noise (very small values of the persistent bars), which makes this method attractive because it can extract the central structural features that allow precise characterization of the system. Importantly, while barcodes may seem intuitive, their statistical analysis is complex.

To this end, we use the definition of persistent entropy by defining the probability of a given topological signature as the length of the persistent bar [2], i.e., $\mathcal{D}\left(\overline{\Gamma}^k\right) = \sum_{i=1}^m \delta_i^k \cdot \log\left(\delta_i^k\right)$.

From this definition, it is also possible to define the relative entropy between two trajectories as

$$S\left(\overline{\Gamma}^{kl}\right) = \sum_i \Delta_{kl}|H_{m,i} \cdot \log\left(\Delta_{kl}|H_{m,i}\right). \tag{5.14}$$

The differences in the trajectories contain topological signatures. If there is a detected entropy in the data, i.e., there is a probability that two trajectories from apparent similar systems diverge, then there exists an intrinsic bias larger than 0, i.e., if the distortion matrix of the measured trajectories is $\mathcal{M}\left(\overline{\Gamma}^{kl}\right) \geq 0$ and the aforementioned persistent intrinsic entropy is larger than zero ($S\left(\overline{\Gamma}^{kl}\right) \geq 0$), then

$$S\left(\overline{\Gamma}^{kl}\right) \geq 0 \text{ and } Bias^{kl}\left(\overline{\Gamma}\right) \geq 0. \tag{5.15}$$

Otherwise, dynamic trajectories and data with a stable structure are generated by systems with an approximately similar inherent structure. Mathematically, this means:

$$\text{if } \mathcal{M}\left(\overline{\Gamma}^{kl}\right) \to 0 \text{ and } S\left(\overline{\Gamma}^{kl}\right) \to 0 \text{ then}$$
$$\mathcal{D}\left(\overline{\Gamma}^{kl}\right) \to 0 \text{ and } Bias^{kl}\left(\overline{\Gamma}\right) \to 0. \tag{5.16}$$

With both of these conditions, we formulate a precise definition of observability:

Definition
Observable systems require low relative persistence of data, i.e., *distorted relative trajectories* $\mathcal{M}\left(\overline{\Gamma}^{kl}\right) \geq 0$. *This implies persistent intrinsic entropy and complexity with a high probability that a customized model* $\hat{f}_k$ *is needed*, i.e., *that* $\hat{f}_l$ *will probably not completely fit the sampled data of k.*

In summary, the goal is to estimate the distortion matrix $\mathcal{M}\left(\overline{\Gamma}^{kl}\right)$ and entropy $S\left(\overline{\Gamma}^{kl}\right)$ of the system to determine whether it can be modeled with a function $\hat{f}$ that accounts for the internal states of the system (see Fig. 5.5).

In this way, we simplify the statistical analysis of persistent bars. In addition, we offer an analysis that is relevant for the quantitative assessment of the relative observability of a system.

5.5.2 Example: Analysis of a Modified Chemotactic System

We test the method presented in the previous section with the modified predator–prey model for chemotactic amoebas, which was presented in Chap. 4. On the basis of the inherent properties of the system and its response to the environment and predator population, the inherent bias (change in element identity) leads to persistent entropy, which can be estimated via Eq. (5.14).

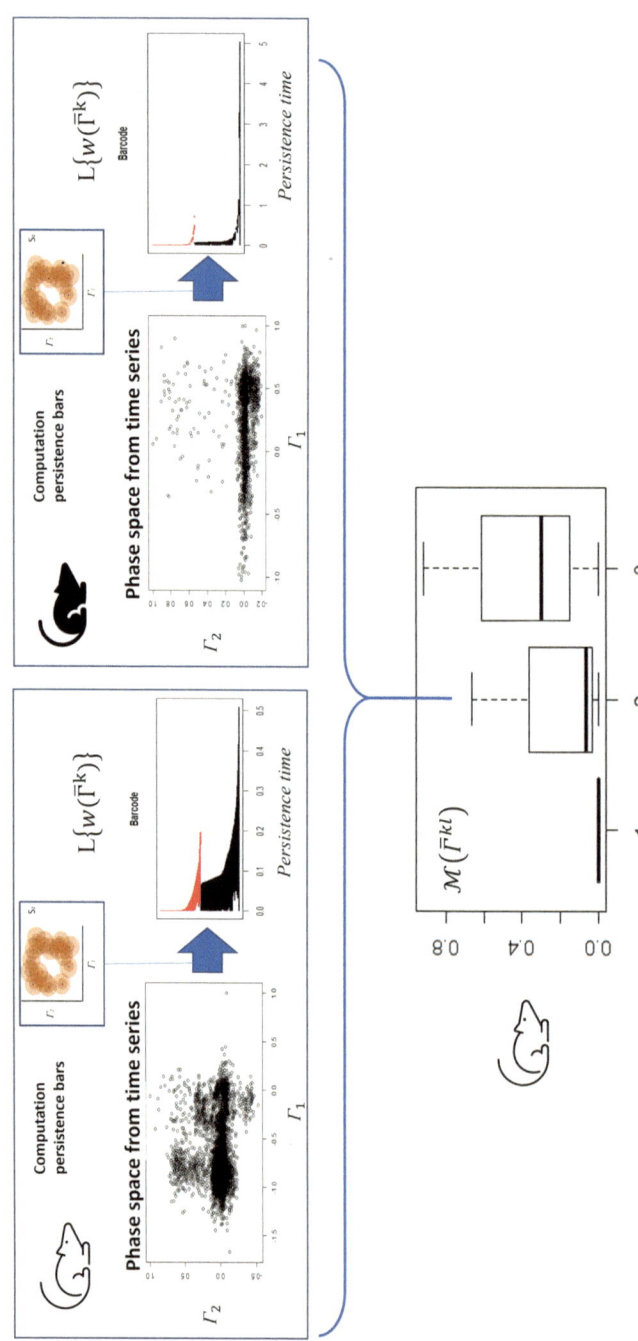

Fig. 5.5 In a phase space, topological invariance is used to analyze the time series of each system or organism. The persistence diagrams are normalized on the basis of the highest value once they have been obtained. As represented by a box plot, the difference between the barcodes of the system (for example, the white mouse) and the other systems is estimated. Reprinted from Diaz Ochoa [7]

For this purpose, we take samples from the time series, extract the topological properties calculated with the persistence bars and calculate the corresponding persistent entropy. An analysis was performed for one population k (for example, predated organisms) relative to the other population l (predators).

The population dynamics of predators and predated organisms with different mechanisms generate three different phase diagrams, which can be characterized by topological persistence and the relative distance of topological signatures.

The matrix $\mathcal{M}^{kl}(\Gamma)$ is represented with heatmaps defined from 0 ($\mathcal{M}^{kl} = 0$) to 1 ($\mathcal{M}^{kl} > 0$) and is compared against the distribution of the relative distance error via boxplots.

Since the switching behavior takes place in a short time period, systems 1 (nonadaptive response) and 2 (adaptive response) can be clustered into the same group, although the error bar of system 1 shows its switching mechanism, suggesting that in this time interval, the mechanisms between systems 1 and 2 are similar, whereas system 3 (coevolving identity) behaves according to different mechanisms.

The persistent entropies confirmed that both systems possess different intrinsic mechanisms and that there is a high relative entropy between the systems, particularly system 3, relative to the other systems (Fig. 5.6).

By detecting the entropy generated by defects in the data and in systems with inherently changing identities, we can conclude that we can use this method to demonstrate the inherent variability between interacting elements of a complex system [7].

Using a similar principle, we want to evaluate the internal degree of distortion of the system in relation to different time periods. This idea offers a way to derive causality from a dynamical system.

5.6 Concepts of Integrated Information

Causation is assessed via a similar notion of entropy. *The quantification of 'causality' is based on the idea that the 'causality' of a variable in relation to another can be measured by how well the variable helps to predict the other. In other words, the variable Y 'causes' the variable X if the ability to predict X is improved by incorporating information about Y in the prediction of X* [28].

This notion requires that there is no loss of information when there is a relationship between two events. Otherwise, the entropy can be measured. The greater the entropy is, the lower the probability that a link will be made between these events.

This concept can easily be transferred to complex mechanistic systems. However, contextual systems are influenced by the environment, implying the need for concepts related to information. Our strategy is to use holistic information concepts.

The goal of this concept, which comes from neuroscience, is to quantify multiple causal influences within a system by using a quantity called integrated information. This serves as a marker of how the nervous system integrates information [20].

The same theory has, however, been used to measure causal influences among elements in systems such as physics, economics, and biology [19]. In biology, this

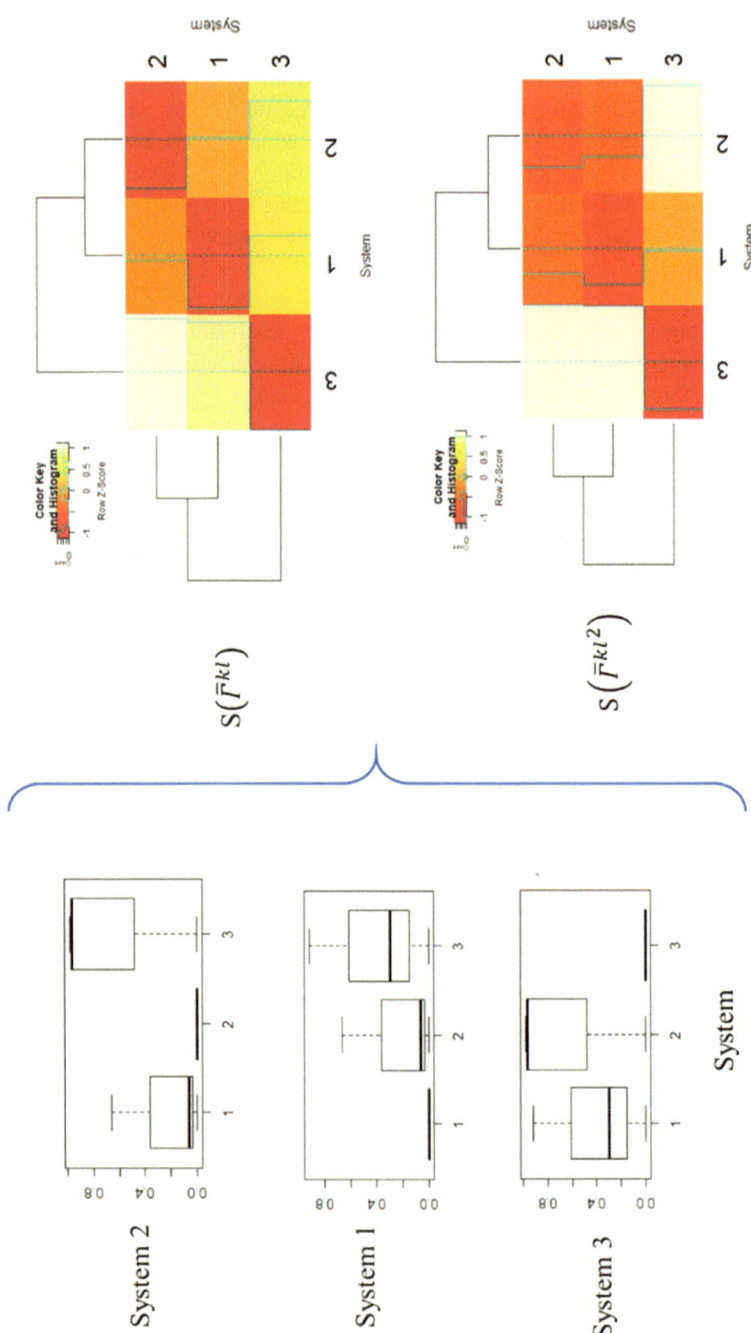

Fig. 5.6 Computed entropy $S\left(\overline{\Gamma}^{kl}\right)$ of the reference system k relative to the other systems l as box plots and sampling of the same entropy as a heatmap. System 1 has a nonadaptive response, system 2 has an adaptive response, and system 3 is a coevolving population switching between both responses. Low variability implies low entropy as well. Reprinted from Diaz Ochoa [7]

concept has been used to understand the evolution of complex organisms on the basis of the relationship between their ability to integrate information and their biological fitness [19], thus linking the concept of biological fitness and concepts of consciousness.

On the basis of our reasoning, the trajectory of a system is calculated through all its microscopic components. The system eventually adapts to the context in which it operates, but this calculation does not always lead to the same result or calculation.

Thus, the implicit structure is equivalent to the structure $\hat{M}$ introduced in the previous chapter, such that the trajectory Γ is the result of the system's computation, i.e., $\mathcal{P}(\hat{M}) = \Gamma$. $\mathcal{P}$ refers to the computation of an output (in this case, the trajectory Γ) on the basis of information stored in the structure $\hat{M}$ (in what follows, $p(x)$ refers to a probability distribution, whereas $\mathcal{P}(\hat{x})$ refers to a system's computation).

Stable mechanisms are associated with constant causal paths (low complexity) that allow estimation of a transfer matrix $A_\tau(X(t - \tau)|X(t))$, whereas changes in intrinsic connectivity led to changes in causal paths (and therefore to changes in measured complexity).

For this purpose, we adapt the "intrinsic cause and effect power Φ" [23] to assess the degree of system autonomy. The theory proposes that fine-grained behavior cannot simply be reduced to interacting elements, thereby providing a measure of system autonomy [20, 23].

Moreover, on the basis of this theory, we adapt the concept of integrated geometrical information Φ_G [24], taking into account all possible connected and disconnected states.

However, instead of analyzing the states as connected/disconnected [23], we assume that we "*observe*" the whole state and evaluate Φ in different time periods, assuming that in such periods, there are connected/disconnected internal states, which cannot be partially or fully accessed and which are encoded by the structure $\hat{M}$.

Geometrical integrated information theory provides basic postulates for the interpretation of intrinsic-causal power in the evaluation of causal relationships from recorded data:

Postulate 1	*The strength of influences is quantified by a minimized difference between fully connected and disconnected models*
Postulate 2	*A disconnected model system satisfies a Markov condition*
Postulate 3	*The difference between a connected and disconnected model is measured by Kullback–Leibler (KL) divergence*

According to these postulates, the causal relationship between different observed quantities can be defined via the notion of distance, according to Oizumi et al. [24]. This notion of distance provides a notion of geometrical information, which is defined as

$$\Phi_G = \min_q D_{KL}\big[p(X, Y)\big]\big[q(X, Y)\big], \tag{5.17}$$

where $X = \{x_1, x_2, \ldots\}$ and $Y = \{y_1, y_2, \ldots\}$ are the past and present states of the system, p is the joint probability function of connected states, q is the joint probability function of disconnected states, and D_{KL} is the Kullback–Leibler entropy, defined as $D_{KL}[p(X, Y)][q(X, Y)] = \sum_{X,Y} p(X, Y) \log \frac{p(X,Y)}{q(X,Y)}$. Accordingly, Φ_G is upper bounded by the mutual information $I(x_{t'}, x_t)$, such that:

$$0 \leq \Phi_G \leq I(x_{t'}, x_t). \tag{5.18}$$

This equation is relevant since it provides a notion of causality that depends on the distance between observed events. It also provides an upper bound that depends on the mutual information of the observed events. Consequently, Φ_G fulfills the conditions for the definition of integrated information [20], i.e., the assessment of an inherent experience that determines a system's autonomy. Therefore, we have different scenarios.

Scenario 1: Low information loss in mechanistic systems; mechanistic responses led to less autonomous and predictable behavior, even in stochastic systems where fluctuations/dissipation led to self-organization

Scenario 2: Loss of information if the system behaves autonomously; in such cases, we assume that systems have some kind of inner experience, or the possibility of calculating and estimating their own trajectories autonomously. This also implies that systems cannot be trivially objectified

Scenario 3: A complete loss of information occurs in the case of entirely disconnected and aleatory[4] systems. In this case, we are referring to fully chaotic systems.

The common probability function of dynamical systems is determined by the trajectory of the system in phase space. Therefore, Eq. (5.17) describes the distance between phase spaces generated by systems with and without connections.

Systems that change their identity and possess a certain degree of autonomy also change their intrinsic states and structures. In Chap. 4, we introduced the idea that system trajectories are calculated mainly via intrinsic information processing, which is based on interconnected structures $\hat{M}$.

Consequently, the computation of a trajectory can be defined as a function p, representing the system's computation, that uses the internal structures $\hat{M}$ to compute the trajectory Γ, i.e., mathematically $P(\hat{M}) = \Gamma$. This notion requires an appropriate method to assess the system's information processing required for such inherent computations.

---

[4] Aleatoric implies complete causal detachment. Stochastic equal distributions could exhibit causal behavior if one assumes that inherent mechanisms lead to uniformly distributed trajectories.

5.7 Integrated Information and Causal Inference

The system's ability to compute its response to the environment is determined by the evaluation of causal events. On the basis of past system information, intrinsic structures can calculate present or future elements of the trajectory if they remain stable. The evaluation of causal relationships and information loss is the core of integrated information theory (IIT).

Mathematically, if a computation of a trajectory $P(\hat{M}) = \Gamma$ exists, then it is possible to compute the transfer matrix $A_\tau(X(t - \tau)|X(t))$ (and in general, the transfer entropy [29]) to estimate future points in the trajectory [28]. More importantly, when the system is observed at different time periods, $P(\hat{M}') = \Gamma'$ exists and has the same transfer matrix, i.e., according to Scenario 1 presented in the previous section:

$$A_\tau(X(t - \tau)|X(t)) = A'_\tau(X(t - \tau)|X(t)) \Leftrightarrow \text{Low information loss} \tag{5.19}$$

On the other hand, autonomous systems that change their implicit states continuously evaluate their environment. This creates a fracture in the intrinsic structure such that $P(\hat{M}') = \Gamma'$ is no longer invariant, violating the transfer matrix symmetry in Eq. (5.19). This condition also implies information loss in the observed system.

After introducing this expression into Eq. (5.18), and on the basis of the result of Oizumi et al. [24], we obtain

$$\Phi_G = \min_q D_{KL}[p(\hat{M}_\tau)][p(\hat{M}_{\tau'})] \leq I(\hat{M}_\tau, \hat{M}_{\tau'}), \tag{5.20}$$

i.e., $\hat{M}_\tau$ and $\hat{M}_{\tau'}$ can be two different structures, where $I(\hat{M}_\tau, \hat{M}_{\tau'})$ represents the mutual information.

According to this definition, the divergence of Φ_G implies a divergence of microstates, which simultaneously measures the degree of autonomy of the system, as the probability distribution $p(\hat{M}_\tau)$ behaves differently in relation to the response to the environment than does the second probability distribution. Note that we compare the system over two different periods of time, which is essentially different from the original method.

Note that this method is useful for tracking possible inherent changes $\hat{M}$. This can only be accomplished if we assume adaptation/evolution of the system. Therefore, instead of analyzing different models that lead to autonomous systems [24], we analyze the coarse-grained state of the system with an underlying structure $\hat{M}$. Again, note that we do not assume any information about the inherent structure.

The analysis of the trajectory provides information about the retention of causal rules and thus about the ultimate autonomy of the system, which can decide on new trajectories by changing its inherent causal relationships (according to Reichenbach's principle [1], as shown in Fig. 5.7).

We cannot analyze connected and disconnected systems on the basis of trajectory alone Γ. However, we can analyze the trajectory structure by estimating the difference

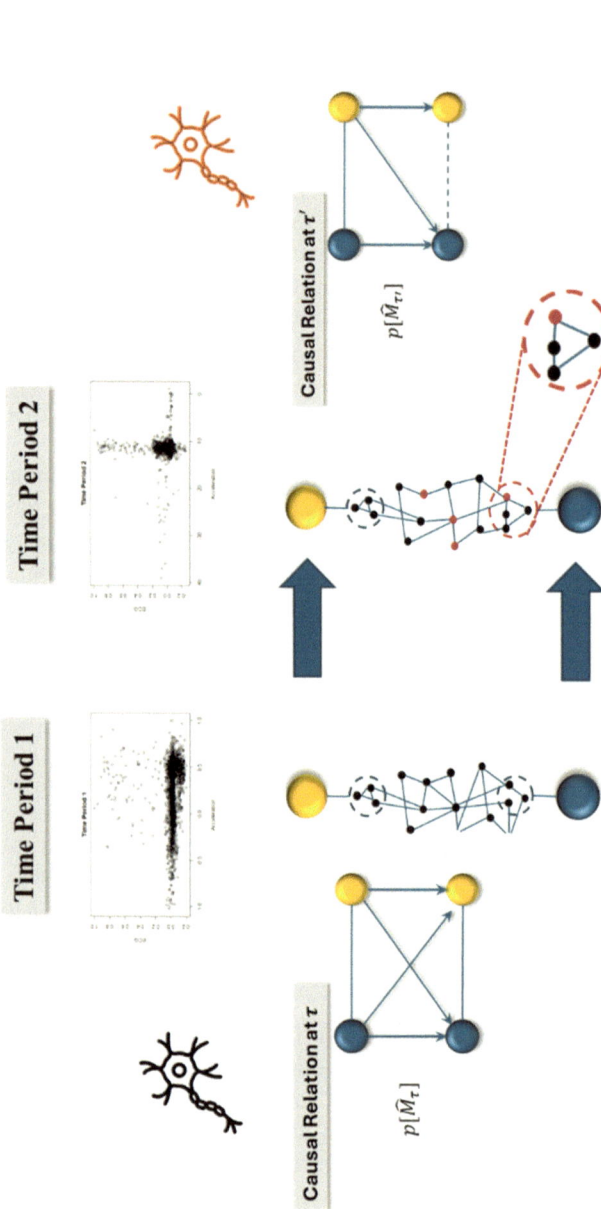

Fig. 5.7 Causal relationships of a system with changing interactions (left and right) lead to a mutation of the system's identity and thus to changes in the observed time series. Instead of defining an interactome,[5] leading to a comprehensive and complete model, we estimate the stability of the coarse graining approach required in the definition of the final observations. We observe here that we make use of Reichenbach's principle by assuming that an inherent structure establishes the causal relationship between two observables

---

[5] https://en.wikipedia.org/wiki/interactome.

in persistence bars $L[\Gamma]$ inside different time periods, as introduced in the previous section [8].

On the basis of the assumption that these distances are implicitly generated by systems with variable connectivity, we compute the Kullback–Leibler distances $L[\Gamma]$ in different time periods as follows:

$$\Phi'_{GP} = \min D_{KL}[L[w_\tau] \| L[w_{\tau'}]] \leq I(\Gamma_\tau, \Gamma_{\tau'}). \qquad (5.21)$$

In this equation, $\Phi_P \sim \Phi'_{GP}$ is the geometrically integrated information based on the persistent topological characteristics of individual trajectories.

According to these postulates, the measurement of the causal relationship between underlying connectivity $\hat{M}_\tau$ and $\hat{M}_{\tau'}$ can be mathematically accounted for as the distance between the topological characteristics of the trajectories.[6] Figure 5.8 shows the workflow of this computation (computations were implemented via the TDA package implemented in R[7]).

In these evaluations, it is impossible to determine if systems have a clear causal structure derived from their underlying mechanics $\hat{M}$. Since systems explore their context/environment in different ways, they should be able to establish different inherent causal relationships to interpret that environment.

This implies that the connectome must also evolve accordingly. On the basis of these postulates, a deviation in the measured probability distribution indicates that a system has changed intrinsically (internal mechanisms and its interaction with the environment), thereby affecting its observation (i.e., the ability to identify mechanisms interlinking different observables, as suggested in Fig. 5.2).

This situation differs from the mere concept of evolving networks because we are tracing not only a constant causal difference in a single system (dynamic and evolving system with dynamic linkages) but also interindividual variations so that some individuals can evolve while other systems possess a stable inherent structure.

Tracking differences between species suggests an inherent degree of autonomy. Therefore, the goal is not to model the uncertainty to obtain a complete model[8] but rather to learn how uncertainties can push systems to act autonomously.

According to these postulates and the concept of mutual information,[9] the upper bound of the mutual information is the join information or join entropy[10] which measures the uncertainty in the set of observations and is defined as

$$H[\Gamma_\tau, \Gamma_{\tau'}] = \sum_{w_\tau, w_{\tau'}} P(\Gamma_\tau, \Gamma_{\tau'}) \log(\Gamma_\tau, \Gamma_{\tau'}) \qquad (5.22)$$

---

[6] Computed in this work using the Kullback Leibler Plugin in R: https://www.rdocumentation.org/packages/entropy/versions/1.3.1/topics/KL.plugin.

[7] https://cran.uni-muenster.de/web/packages/TDA/vignettes/article.pdf.

[8] It is the goal of several modern modeling technologies.

[9] https://en.wikipedia.org/wiki/Mutual_information.

[10] https://en.wikipedia.org/wiki/Joint_entropy.

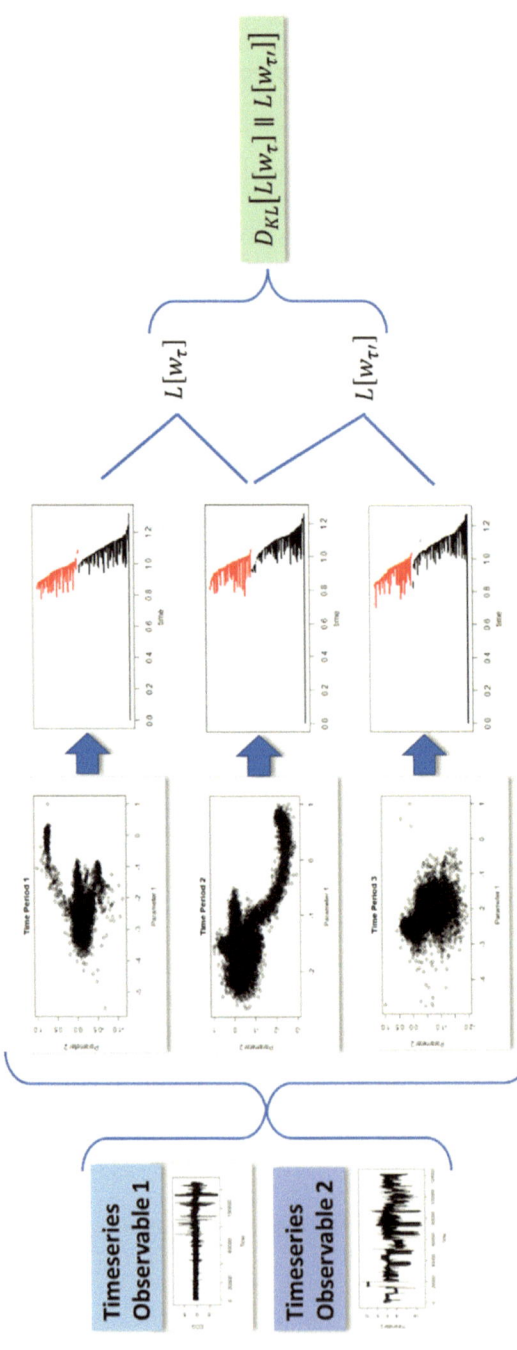

Fig. 5.8 This workflow is based on the computation of persistent bars derived from time series and takes into consideration the construction of clouds of points derived from time series extracted at different intervals, the computation of persistence bars, and the final calculation of the Kullback–Leibler entropy for estimation of Φ_{GP}. Reprinted from Diaz Ochoa [9], Copyright 2012, with permission from Elsevier

where $P(\Gamma_\tau, \Gamma_{\tau'})$ is the joint probability. Accordingly, Eq. (5.23) takes the following form:

$$0 \leq \Phi_{GP} \leq H[\Gamma_\tau, \Gamma_{\tau'}]. \tag{5.23}$$

If $\Phi_{GP} \to 0$, then the connectivity remains invariant and constrained into a single manifold. Otherwise, the structure of the probability distribution diverges. Therefore, when $\Phi_{GP} \to H[\Gamma_\tau, \Gamma_{\tau'}]$, this implies that the derived causal relationships tend to have the same degree of observed uncertainty, eventually leading to extremely variable and aleatory systems (see Fig. 5.3). The interpretation of this expression and its practical application can be found in the next section.

5.8   The Φ_S Complexity Measure

In Sects. 5.5 and 5.7, we presented methods to measure the relative variability and causal inference of both systems. As an observer, I should be confident in reducing every observed system to one object.

Only well-defined objects can be measured with high accuracy and independently of other observers, which means that objects can be described universally.

This statement is relevant because only universal observations and descriptions of well-defined objects can be used in defining fundamental theories and formulating predictive models.

Available information about the observed system must be complete to file as an objective observation. However, if the observed system is autonomous (and behaves subjectively), then there is information loss.

The less objective the system description is, the greater the degree of information loss. Thus, a small-world definition requires small or no-information loss. On the other hand, large-world systems continuously adapt to context and the environment and experience persistent information loss.

In this section, both measurements are combined into a single quantity. In doing so, we aim to provide a kind of chartography that provides the "observability coordinates" of a complex system on the basis of the assumption that a complex system is essentially autonomous and individual (variable) [9].

The Φ_S *complexity* provides an answer to the following questions:

(i) *Can the underlying mechanisms and theory be extrapolated to another organism or population? and*
(ii) *Do these mechanisms obey permanent causal relationships?*

If both questions can be answered with "yes", there is a basis for universal theories as well as models that can be extrapolated to any other system or organism. Otherwise, tailor-made models and constant observations are needed.

According to this representation, this complexity measurement is not a scalar (like entropy or any other complexity measurement on the basis of entropy) but rather a vector for each observed organism or system ordering.

This vector depends on the amount of integrated geometric information Φ_{GP} (causal analysis) and the mean amount of persistent entropy $S(\Gamma^{kl})$ (according to Eq. 5.14), i.e., $S(\Gamma^k) = \frac{\sum_l S(\Gamma^{kl})}{N_k}$, where N_k is the total amount of other reference organisms. Thus, the Φ_S complexity for each organism k is defined as the vector defined using the entropies introduced in Eqs. (5.14) and (5.16):

$$\Phi_S^k = \Phi_S\{\Phi_{GP}(\Gamma^k), S(\Gamma^k)\} \tag{5.24}$$

The representation of this vector and the distances measured on the main axes Φ_{GP} and $S(\Gamma^k)$ and the relative distances to low entropy values provide the coordinates to assess observability/autonomy in a population of similar complex systems.

The Φ_S complexity can be represented as a vector in a Cartesian system (see Fig. 5.9). Importantly, defining thresholds that represent observability boundaries is more important than complexity space: for deterministic systems, these values provide the boundaries of areas of high or low observability, depending on how variable the systems are.

This diagram essentially is a space for the vectors Φ_S^k (i.e., the individual complexity for each system's element) and represents the degree of disorder in persistent patterns derived from the measured paths, where the x axis, $S(\Gamma^k)$, represents the relative degree of disorder of these paths (degree of individual behavior) and the y axis, $\Phi_{GP}(\Gamma^k)$, represents the degree of autonomy (see Fig. 5.9) (in what follows, we will simply refer to this complexity as the Φ_S complexity, i.e., ignore the index k for the measurement of this complexity for each element).

It is important to ask if the information this graph provides is skewed by variation and noise. Equation (5.24), which was used to define this diagram, essentially depends on the persistent bars, which are particularly capable of detecting and filtering noise (see Fig. 5.4); the short persistent bars in the diagram therefore essentially represent noise and are filtered in the final evaluation.

A system's observability and inherent autonomy cannot be defined without threshold values. A rather simple and comprehensive analysis takes extreme values into account, e.g., if the system is observable ($\Phi_S \to 0$, low complexity), stable inherent causal structures are responsible for mechanisms leading to model identification. The larger this complexity value is, the lower the system's observability. The following are more exact conditions for assessing system observability:

i. *Assuming that half of the persistent patterns are mutually divergent, the entropy takes the following value $S(\Gamma^{kl}) \approx 0.5/N_k$, where N_k is the number of observed elements. As a result of this approximation, a threshold can be defined to determine whether the states of the system differ from each other: below this value, persistent patterns are usually not divergent (similar mechanisms in different systems, region a in Fig. 5.9); otherwise, the patterns diverge, indicating high variability between systems (region b, Fig. 5.9).*

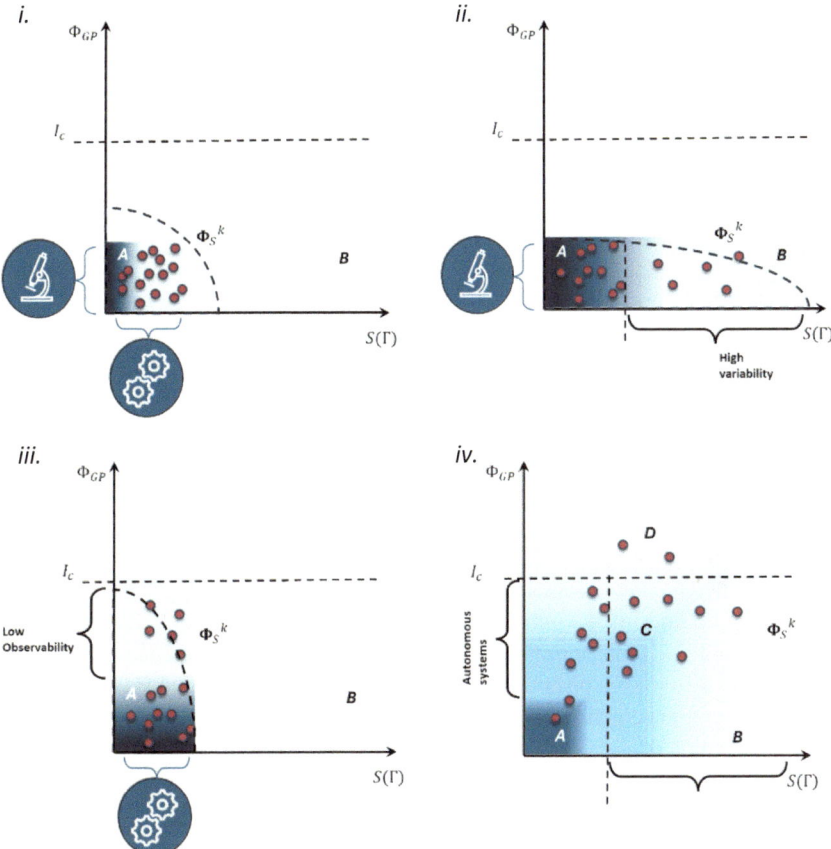

Fig. 5.9 The vector space can be used to represent different forms of Φ_S complexity. Systems with low complexity (low observability and low dispersion) consist of short vectors that accumulate at the corner of the diagram (**i**). As a result of system complexity, the length of the vectors increases. The systems shown in **ii** and **iii** can either have high variability and still be observable, or they can be unobservable with minimal variability. In addition, systems with high aleatory values that are not observable (i.e., autonomous systems) are associated with large vectors that propagate away from the corner of the graph (**iv**)

ii. *When Φ_{GP} is below the mean joint entropy and the mean joint entropy is relatively low, the system behaves mechanically; otherwise, there are several phase transitions in the system, and it is likely that it will behave autonomously due to its information processing (region c, Fig. 5.9) or aleatory (region d, Fig. 5.9).*

The autonomy of the system is not valued in the same way as the integrated information is, as we do not have access to the system's structure. Indirectly, however, Φ_S indicates that the trajectories of the system are undergoing phase changes that may not be attributed to autonomous behavior when $0 < \Phi_{GP} < H[\Gamma_\tau, \Gamma_{\tau'}]$. In addition, interindividual variability suggests that these phase changes are both individual and autonomous.

5.9   Applications of the Φ_S Complexity Measure

Throughout this section, we discuss how to interpret the complexity measure, which is defined as the vector Φ_S in Eq. (5.24). By doing so, we will be able to gain a deeper understanding of the observability of complex systems. As an example, we examined the data in the M-Health database[11] [3].

This database is interesting because it is obtained from people who wear smart devices that track training and heart response in real situations and is an ideal method to test the theory presented in this example, since the analysis of data obtained in real conditions contains information about the body's response to real external conditions.

To assess a participant's response to exercise, measurements were taken on the person's devices, which measure acceleration along three axes and other physiological parameters, such as the ECG. This data was collected as part of a project to develop methods for integrating mHealth data.

During the experiment, participants wore a chest-mounted sensor that allows for both 2-lead ECG measurements and basic heart monitoring. Sensors measuring acceleration (in m/s^2) are placed on the chest, right wrist, and left ankle of the subject and are attached via elastic straps.

The aim of this study was to measure the impact of exercise in real-world settings outside the laboratory. In addition, the basic concept behind this study is to take a systemic approach in describing the training response; therefore, the heart rate control system was designed as a complex network with nonlinear feedforward and feedback inputs.

These data show chaotic and nonlinear dynamics as a result of interactions between physiological oscillators, functional state changes, and noise [10, 32].

Although this nonlinear property exists, we assume that a simple model can be developed on the basis of the correlation between the two states $A \leftrightarrow B$, i.e., $A(x, y, z) = f_1(B)$, and $B = f_2(A(x, y, z))$, where $A(x, y, z)$ is the acceleration measured in three different axes and where is an electrocardiogram of the heart response (Fig. 5.10a).

As a result of this simple model, there is a high probability that different individuals in a small population have similar causal relationships in their responses to exercise, i.e., the heart response and physical activity are simple and have a mechanistic relationship.

This 3D representation can be useful in understanding the meaning of the present analysis: when the complexity measure "condenses" inside the blue bubble (i.e., vectors with a small length), causal relationships are stable, indicating that this system

---

[11] https://archive.ics.uci.edu/ml/datasets/MHEALTH+Dataset.

A.

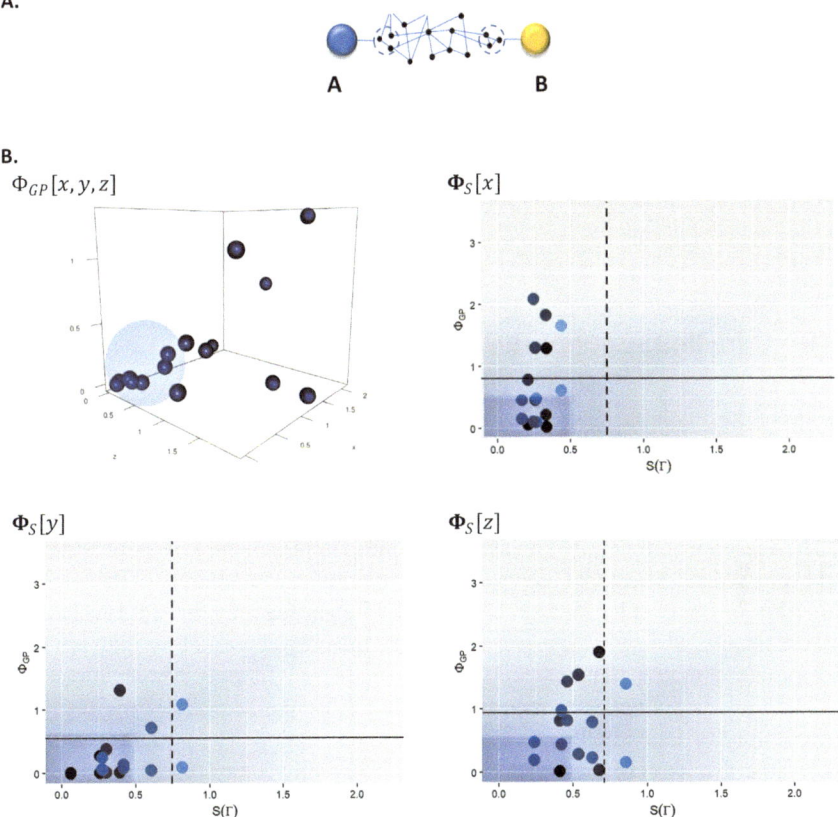

B.

Fig. 5.10 According to **a**, physical activity (measured as acceleration **a**) is correlated with the cardiac response (measured as **b**). Figure **b** shows the computation of the $\Phi_{GP}[x, y, z]$ and Φ_S diagrams for each acceleration axis. A low causal entropy is required to induce or deduce a model, but the results of a first inspection of this entropy along the three axes indicate that only a few individuals possess this low entropy (**b**, 3D diagram). Reprinted from Diaz Ochoa [9], Copyright 2012, with permission from Elsevier

can be modeled; that is, the recognized patterns can serve as features for inductive models (for example, using artificial intelligence).

Furthermore, "informational evaporation" occurs near the condensation region, which indicates that the system is autonomous. The system is either highly autonomous or aleatory if there is a large amount of "informational evaporation". These results indicate that there is no chaotic or aleatory behavior, as expected.

Consequently, we found a de facto correlation between cardiac response and physical activity. The close agreement with the blue area confirms the feasibility of identifying both inductive and deductive models or a combination of both. Nevertheless, there are some strong deviations that indicate that coupling is more complex than just chaotic (Fig. 5.10b).

Some individuals have different patterns of movement on the x-axis, suggesting an autonomous response to movement. In addition, there is a high degree of interindividual variability, so that only a few individuals can be modeled correctly with a universal model.

Since only a few individuals lie within the blue square, which is reserved for almost all mechanistic systems, we conclude that any modeling effort will involve a high degree of uncertainty and that a common calibrated model might be feasible for only a few individuals.[12]

5.10 Concluding Remarks

We introduced complexity measurement to assess the observability of a complex system relative to other similar systems, taking into account both the inherent variability and autonomy that limit the observability of the system and especially the estimation of causal pathways.

The degree of complexity is measured for one system element in relation to another and expressed in a space that represents the variability and autonomy of that element. This form of complexity measurement has two main features: first, it is relative to other elements, and second, it is a vector-like quantity representing the variability and autonomy of that element.

Like the navigating of a map, this graph can be used to assess the degree of observability of individual organisms within a population. The system is probably observable if both entropies tend to be zero. Once this is ensured, it is possible to evaluate whether the inherent mechanisms of the system can be mapped out by a black-box (or white-box) model.

This approach can be used to determine whether a theoretical approach is the most appropriate way to understand the system. Additionally, with this method, we can determine how reliable a mathematical model of a system is by identifying its relevant features from the data.

A discussion and speculation about the implications of the concept of elastic states and the methodology introduced here are presented in the following chapter.

References

1. Allen J-MA, Barrett J, Horsman DC, Lee CM, Spekkens RW (2017) Quantum common causes and quantum causal models. Phys Rev X 7:031021. https://doi.org/10.1103/PhysRevX.7.031021
2. Atienza N, Escudero LM, Jimenez MJ, Soriano-Trigueros M (2019) Persistent entropy: a scale-invariant topological statistic for analyzing cell arrangements. arXiv 190206467 Cs

---

[12] All the reported computations are reported in the following repository: https://github.com/2001Odisea/Phi_S_Diag_Persistent.

3. Banos O, Garcia R, Holgado-Terriza JA, Damas M, Pomares H, Rojas I, Saez A, Villalonga C (2014) mHealthDroid: a novel framework for agile development of mobile health applications. In: Pecchia L, Chen LL, Nugent C, Bravo J (eds) Ambient assisted living and daily activities. Lecture notes in computer science. Springer International Publishing, Cham, pp 91–98. https://doi.org/10.1007/978-3-319-13105-4_14

4. Chevyrev I, Nanda V, Oberhauser H (2020) Persistence paths and signature features in topological data analysis. IEEE Trans Pattern Anal Mach Intell 42:192–202. https://doi.org/10.1109/TPAMI.2018.2885516

5. de Amorim RC, Hennig C (2015) Recovering the number of clusters in data sets with noise features using feature rescaling factors. Inf Sci 324:126–145. https://doi.org/10.1016/j.ins.2015.06.039

6. Diaz Ochoa JG (2018) Elastic multi-scale mechanisms: computation and biological evolution. J Mol Evol 86:47–57. https://doi.org/10.1007/s00239-017-9823-7

7. Diaz Ochoa JG (2020a) Observability of complex systems by means of relative distances between homological groups. Front Phys 8. https://doi.org/10.3389/fphy.2020.465982

8. Diaz Ochoa JG (2020b) Observability of complex systems by means of relative distances between homological groups. Front Phys 8:503. https://doi.org/10.3389/fphy.2020.465982

9. Diaz Ochoa JG (2023) A unified method for assessing the observability of dynamic complex systems. Comput Biol Med 160:107012. https://doi.org/10.1016/j.compbiomed.2023.107012

10. Dimitriev D, Saperova EV, Dimitriev A, Karpenko Y (2020) Recurrence quantification analysis of heart rate during mental arithmetic stress in young females. Front Physiol 11:40. https://doi.org/10.3389/fphys.2020.00040

11. Du D (2014) Contributions to persistence theory. arXiv 12103092 Cs Math

12. Edelsbrunner H, Letscher D, Zomorodian A (2000) Topological persistence and simplification. In: Proceedings 41st annual symposium on foundations of computer science, pp 454–463. https://doi.org/10.1109/SFCS.2000.892133

13. Emrani S, Chintakunta H, Krim H (2014) Real time detection of harmonic structure: a case for topological signal analysis. In: 2014 IEEE international conference on acoustics, speech and signal processing (ICASSP), pp 3445–3449. https://doi.org/10.1109/ICASSP.2014.6854240

14. Fasy BT, Lecci F, Rinaldo A, Wasserman L, Balakrishnan S, Singh A (2014) Confidence sets for persistence diagrams. Ann Stat 42:2301–2339. https://doi.org/10.1214/14-AOS1252

15. Ghrist R (2007) Barcodes: the persistent topology of data

16. Ghrist R (2008) Barcodes: the persistent topology of data. Bull Am Math Soc 45:61–75. https://doi.org/10.1090/S0273-0979-07-01191-3

17. Gowdridge T, Dervilis N, Worden K (2022) On topological data analysis for structural dynamics: an introduction to persistent homology. https://doi.org/10.48550/arXiv.2209.05134

18. Hastie T, Tibshirani R, Friedman J (2009) The elements of statistical learning: data mining, inference, and prediction, 2nd edn, corr. 9th printing 2017. Springer, New York, NY

19. Joshi NJ, Tononi G, Koch C (2013) The minimal complexity of adapting agents increases with fitness. PLoS Comput Biol 9:e1003111. https://doi.org/10.1371/journal.pcbi.1003111

20. Koch C (2019) The feeling of life itself: why consciousness is widespread but can't be computed, illustrated edn. The MIT Press, Cambridge, MA

21. Liu Y-Y, Slotine J-J, Barabási A-L (2013) Observability of complex systems. Proc Natl Acad Sci 110:2460–2465. https://doi.org/10.1073/pnas.1215508110

22. Mao X, Shang P (2017) Transfer entropy between multivariate time series. Commun Nonlinear Sci Numer Simul 47:338–347. https://doi.org/10.1016/j.cnsns.2016.12.008

23. Marshall W, Kim H, Walker SI, Tononi G, Albantakis L (2017) How causal analysis can reveal autonomy in models of biological systems. Philos Trans R Soc Math Phys Eng Sci 375:20160358. https://doi.org/10.1098/rsta.2016.0358

24. Oizumi M, Tsuchiya N, Amari S (2016) Unified framework for information integration based on information geometry. Proc Natl Acad Sci U S A 113:14817–14822. https://doi.org/10.1073/pnas.1603583113

25. Pereira CMM, de Mello RF (2015) Persistent homology for time series and spatial data clustering. Expert Syst Appl 42:6026–6038. https://doi.org/10.1016/j.eswa.2015.04.010

26. Pienaar J (2017) Causality in the quantum world. Physics 10:86. https://doi.org/10.1103/Phy sRevX.7.031021
27. Pun CS, Xia K, Lee SX (2018) Persistent-homology-based machine learning and its applications—a survey. arXiv 181100252 Math
28. Razak FA, Jensen HJ (2014) Quantifying 'causality' in complex systems: understanding transfer entropy. PLoS ONE 9:e99462. https://doi.org/10.1371/journal.pone.0099462
29. Schreiber T (2000) Measuring information transfer. Phys Rev Lett 85:461–464. https://doi.org/10.1103/PhysRevLett.85.461
30. Siegenfeld AF, Bar-Yam Y (2020) An introduction to complex systems science and its applications. Complexity 2020:6105872. https://doi.org/10.1155/2020/6105872
31. Takens F (1981) Detecting strange attractors in turbulence. In: Rand D, Young L-S (eds) Dynamical systems and turbulence, Warwick 1980. Lecture notes in mathematics. Springer Berlin Heidelberg, pp 366–381
32. Voss A, Schulz S, Schroeder R, Baumert M, Caminal P (2009) Methods derived from nonlinear dynamics for analysing heart rate variability. Philos Trans R Soc Math Phys Eng Sci 367:277–296. https://doi.org/10.1098/rsta.2008.0232

Chapter 6
Life and Completeness in Complex Systems

"Ignoramus" and "Ignorabimus"
Emil du Bois-Reymond

Keywords Incompleteness · Decision making · Basal cognition · Systems theory · Control theory · Consciousness · Artificial intelligence · Ethics

In Chaps. 4 and 5, we have shown the effect of the interaction of elements with elastic identities in complex systems. In these chapters, we discuss the difficulty of clearly defining an identity in a multiscale system and how to approach this problem mathematically from the perspective of causal inference.

According to this theory, complex systems are generally not individual labyrinths (with a clear single architecture) but a labyrinth within labyrinths.

In this last chapter, we argue that living systems aim for low incompleteness and "seek" an agreement,[1] leading to well-defined contexts and stable (nonelastic) states. To use the metaphor of the labyrinth, maintaining a low level of incompleteness is tantamount to finding a maze and its exit among a series of possible labyrinths.

This is only possible if the close interrelationship between individual interacting elements and their context is taken into account. This also implies that life is unlikely to be summed up in a single equation or statistical description.

We review the standard theories of complex systems and argue that there are natural limits to the mathematical representation of complex systems. We also speculate on how life could play a much more fundamental role than we thought and think about the next steps and open questions.

---

[1] According to the Cambridge dictionary, this is a form of harmony, in which different parts seek a common agreement: https://dictionary.cambridge.org/dictionary/english/harmony.

J. G. Diaz Ochoa, *Complexity Measurements and Causation for Dynamic Complex Systems*, Understanding Complex Systems, https://doi.org/10.1007/978-3-031-84709-7_6

6.1   The Standard Theory of Systems Theory: An Overview

How can we extract and understand the basic structure of a complex system? The answer is to systematically disassemble the system, isolate its basic parts, and then try to understand their couplings and interactions with their environment. In this way, the notions of causality, either punctual causality or the extraction of causal properties, can help us understand the underlying mechanisms. We can then extract the laws that govern their behavior and evolution (see Chap. 2).

6.1.1   "Small World" and Systems Theory

In Chap. 3, from the concept of small worlds, it is possible to understand the construction of complex systems by dissecting and extracting the simplest interconnected elements, analyzing their interactions and their interlinking in an ever-growing hierarchy of small worlds on different scales, where simple interaction principles can be precisely empirically characterized and then mathematically represented.

Here, the small world means that fundamental interacting elements can be isolated, or similarly, that interacting large worlds can be dissected into single interconnected elements, such as from the understanding of relatively simple microscopic elements, it is possible to discover how complex systems are built up (for example, the function and reaction to toxic substances of the liver; see Fig. 6.1).

As noted previously, such an approach is equivalent to the notion of an "ensemble" in statistical physics [76]. This concept helps understand the behavior of several particles in an experimental configuration.

Thus, from fully isolated systems called microcanonical ensembles (without flow of energy and matter), it is possible to understand basic phenomena that lead us to the theoretical description of many complex ensembles (grand canonical ensembles) in which there is a flow of particles and energy from an external reservoir (see Chap. 3).

A system of interacting elements generates an average control function on the basis of the physical constraints over a microscopic interaction. Here, it is possible to recognize a basic principle: while the control function tells microscopic systems how to interact, the single interconnected elements tell us how to define the whole control function (see final section, Chap. 3).

This control function has the same character as the field in field theory in physics; for this reason, physics is often used as a discipline to derive fundamental principles of complex systems.

This implies that after understanding the mechanisms of the microscopic world ("small world") combined with the statistical understanding of the system, it is therefore possible to derive fundamental axioms related to the system's causal property, allowing the derivation of primordial and universal laws.

In summary, the standard model in systems theory aims at understanding any complex system being a natural consequence of an ever-growing complexity that

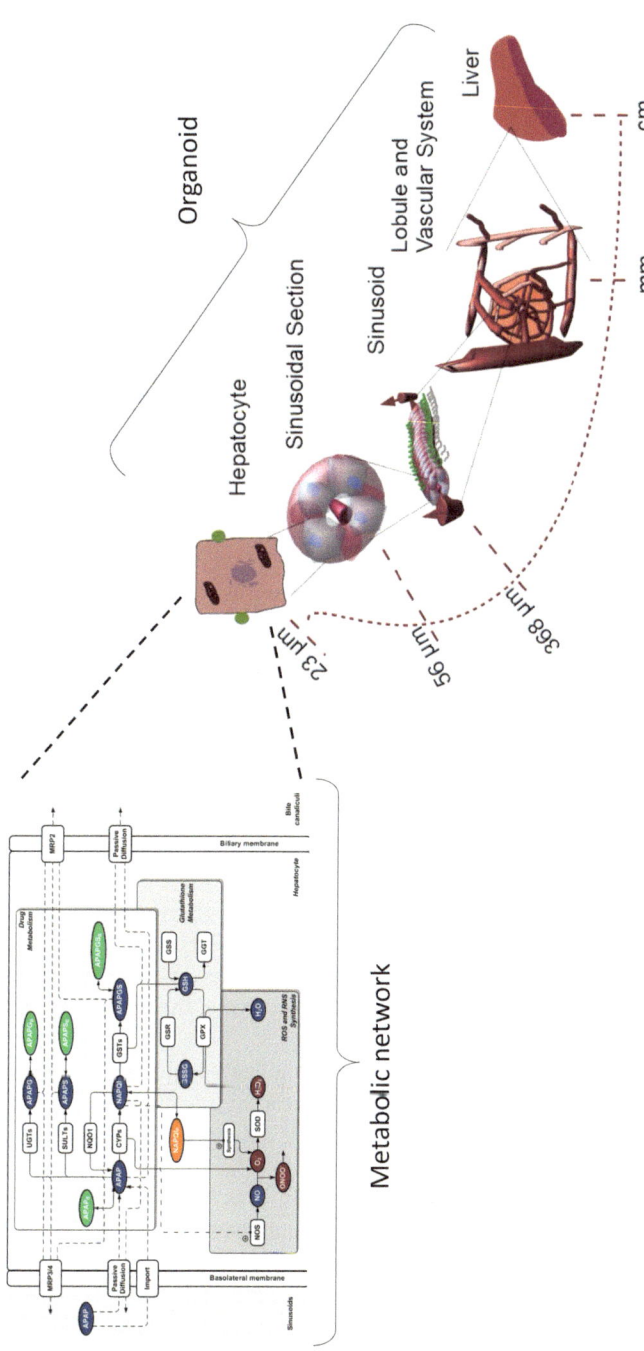

Fig. 6.1 To study complex systems scientifically, each system within the systems must be dissected. For example, from the understanding of a metabolic network, it is possible to understand how hepatocytes transform a substance (in this case, APAP); hepatocytes can then be coupled to organelles representing elemental and anatomical liver units, from hepatocytes to liver organelles. In this way, it is possible to study the effects of hepatotoxic substances at different concentrations on the liver. Both figures are reprinted from Diaz Ochoa et al. [21]

starts from simple principles by constructing a model f capable of predicting an expected system's output O, starting from input data or parameters I, i.e., $f : I \rightarrow O$. This mathematical function f is a model that represents the intricate interrelations between simple and complex elements, as well as complex flows of matter, energy, and information.

With this in mind, systems theory (as a type of field theory of complex systems) aims to extract all the fundamental interactions and principles that control and synthesize natural systems. In extreme approaches, this concept assumes that the principles of complex systems (including living systems) are universal and that complex systems are therefore deterministic.

6.1.2   Systems Theory and Networks

According to the previous section, complex systems can be governed by fundamental principles that can be organized axiomatically. Here, the concept of emergence plays a central role, in which the coupling between microscopic interactions, fluctuations and dissipation guides the construction of increasingly complex structures [30].

As shown in Chap. 3, when discussing emergence, we use the concept of causal property. Similarly, instead of defining causality up and down (which is a concept conventionally used in complex science, see, e.g., Capra and Luisi [10]), we define a control function that is constrained by single elements, whereas the control function constrains the individual states of the interacting elements (see Chap. 3, Fig. 3.11).

Thus, the collective interactions in combination with (fine-tuned) parameters lead to a system that is put into a critical state that spontaneously generates order and structure.

As in statistical physics, the smallest elements determine the function of the system, whereas the overall system determines how the individual elements should behave. In systems theory, this argument is central, especially against genetic determinism as the primary cause of an organism's function and phenotype [10, 54].

Similarly, as we have shown in Chap. 3, the connectivity and feedback loops of different elements influence collective behavior. Therefore, the dynamics of the system determine the dynamics of the local element.

A classic example of this is the feed loop of a ball regulator controlling the flow in the pipeline on the basis of the rotation of its ball regulator, which rotates at an angular velocity that depends on the flow in the pipeline and is simultaneously coupled to a valve that regulates the flow rate (see Fig. 6.2).

As a result, network-like structures have been defined as systems with interconnected elements, including electronic circuits, protein networks, and cellular signal transduction. This vision is inspired by an engineering paradigm: understanding control functions improves the understanding of natural systems. Once the "natural" control functions are identified, it should be possible to control natural systems at any scale.

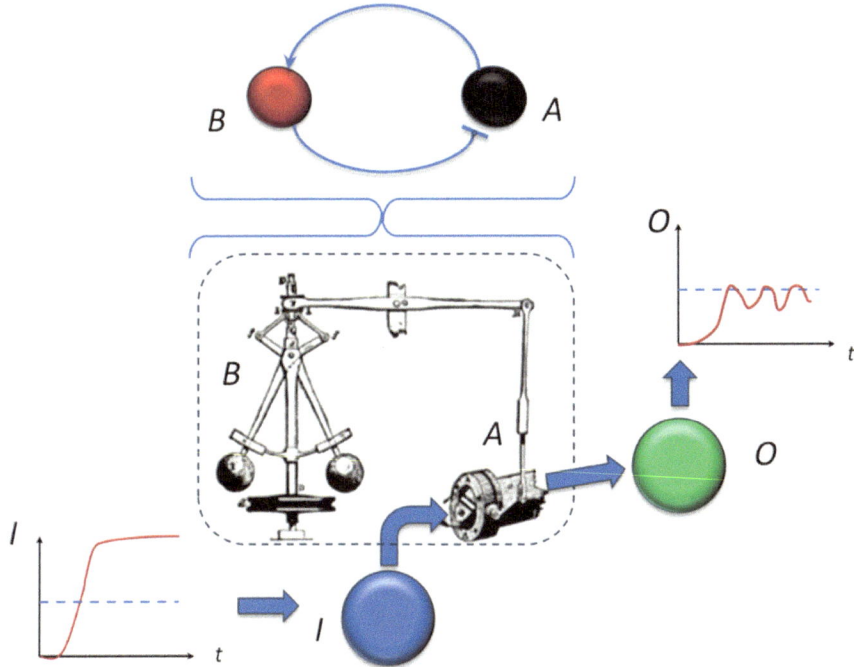

Fig. 6.2 The centrifugal ball regulator serves as a self-regulating principle. This type of feedforward system has become a paradigm for understanding natural systems. Here, the isolation of the single "small world" principles (centrifugal force and action on a valve) allows the self-regulation of a complex steam machine. *Source* Centrifugal ball governor: Wikipedia

Figure 6.1 shows the metabolism of hepatocytes, coupled with a multiagent system and a physiologically based pharmacokinetic (PBPK) model, in the distribution of a substance throughout the body, thus allowing the modeling of the uptake and absorption of a substance such as paracetamol [21].

6.1.3   Systems Theory and Life

If one follows this line of reasoning, then life is the byproduct of the interrelationships of complex systems. Similarly, cognition and awareness are also final byproducts of this complexity.

They represent the apex of the evolution of biologically complex systems, with human consciousness being the end result of the complex development of the nervous system and brain. On the basis of this knowledge, in recent years, there has been rapid development in the field of neuroscience, stimulating the development of artificial systems that simulate brain function, which has led to rapid development in the field of artificial intelligence, as well as the establishment of specific criteria to define how artificial systems can become conscious [42].

Notably, consciousness is more likely to have a function in systems with a high degree of organization (e.g., to navigate a complex environment and meet the law of requirements for complex systems [69]).

However, complex systems can also evolve without cognitive ability. The most important feature of a complex system is its ability to regulate its energy consumption and replication capability. In principle, these properties do not require cognitive skills, and simple ways to address neg entropy are sufficient to maintain a complex system.

In particular, the lack of cognitive ability and the ease of response to resource consumption and replication are the basic assumptions for successful modeling of multiagent populations. This has been shown in the game of life, where evolution is only determined by initial conditions [26]. Thus, any cognitive capability could just be an artifact.

According to systems theory, life, as an intricate form of a complex system, plays a pivotal role in the construction and coordination of the great complexity of social organisms. It regulates both energy and information flow. Life is essentially the result of either a growing complex interaction between fundamental elements or a newly emerging phenomenon that can be controlled and replicated.

However, such an approach has led to some marginal advances in the actual understanding and control of complex open systems [69].

Engineering designs and controlled complex systems by isolating and characterizing single interacting elements. The hope is that by understanding isolated systems, it will be possible to understand real open systems. Nevertheless, physics, engineering, and systems theory have limits when taking whole systems in different contexts into consideration.

In engineering, technical systems perform functions from their creator's perspective but not from other systems' perspectives.

6.1.4 System Theory Versus System Function

Naturally, complex systems have functions from different perspectives. Seed husks, for example, are used not only to transport and protect seeds (single function) but also to feed animals and build bird nests. In addition, there is an inexplicable kind of beauty that goes beyond the mathematical symmetries in physics [70], such as the diversity of colors in animals or plants and the diversity of shapes and forms.

Beauty has often been considered irrelevant from a scientific perspective—in part because it is intrinsic and subjective [78]—but has recently been recognized as a relevant intrinsic aspect in the function of complex systems [36].

This concept of *system function* in complex systems (and in particular of *biological function* [22]) is much more evasive than the concept of causal potency that natural sciences (especially physics) and engineers visualize when understanding a complex system [68].

As a rule, system function is also considered in molecular biology and bioengineering, for example, in plant development [52].

However, the function of the system is often ignored, as it is not well understood as a whole, such as in ecosystems [1]. Perhaps the serious consequence of ignoring these features is that science and technology provide techniques to disrupt complex systems rather than control them.

This can be especially problematic when technical systems are scaled up, which may become a life-threatening condition for all humanity[2] (see Sect. 6.3).

Technical systems can also have a function from a biological point of view. Let us use vehicles, for example. When designing a motor vehicle, the goal is to transport people in a safe way (and eventually provide some fun to drive).

However, from the perspective of a natural system, the purpose of a car is much greater than being one of the largest predators in the world [27]. Perhaps motor vehicles serve as a type of regulation for the population. Or should cars be a niche for some species? For example, some pioneer plants use motor vehicles as vectors to propagate their seeds [86]. In this case, cars are forms of technology that do not have any needs but can be passively utilized by animals, plants, and other organisms.

Therefore, it is appropriate to assume that the basic understanding of complex systems is still open. In addition, many processes are probably not emergent but fundamental. We discuss these aspects in the next section.

6.2 Decision-Making and System Function

6.2.1 Limitations of Systems Theory

Why are complex systems still so elusive, and why does the world seem to be on the brink of chaos rather than in a controlled state? Even basic models based on transformer-like architectures (implemented in large language models) cannot be generalized and used outside of the box, and in many cases, such models undergo further specialized training with specialized datasets to improve their ability to predict specific properties and functions, such as protein structure or gene expression [23].

Any description based on simple axioms, no matter how complex, requires absolute and objective methods to describe interacting elements across different scales. As we have observed in Chap. 3, causal factors determine system dynamics, with intrinsic properties that can be characterized physically.

Since any complex system originates from single principles leading to causal factors with perhaps perfect symmetry, it must be possible to understand the whole universe via single and elegant principles. One example is the search for a unified field theory as the final origin of the whole complexity in the universe [57]

---

[2] This also applies to politics, in which authoritarian regimes hinder democracies and societies and exert their influence through autocratic powers. The goal of such measures in a system is to chronically disrupt and weaken societies which are then open to dictatorships.

However, in Chaps. 4 and 5, we have shown that open complex systems are generally incomplete, i.e., they cannot be described in a simple way, since the context of the system cannot be defined in absolute terms in multiscale systems. This is because localized ecological systems are known to abruptly and irreversibly switch from one state to another when forced above their critical thresholds [3].

Therefore, the control functions necessary for the definition of the right mathematical models cannot be trivially computed.

Nonetheless, interaction with the environment and the selection of its capabilities induce primitive cognition that supports interacting elements in decision-making [6] and the selection of its function.

Changing environments force living organisms to continuously strive to maintain decidability or relative completeness despite constant changes in the environment. This limits our ability to identify models in systems biology: while the mechanism can be valid under certain circumstances, the continuous assimilation and accommodation of organisms challenge the completeness of these mechanisms [19].

This implies a persistent distortion of the interaction mechanisms. Thus, "higher scales" in biology close the loop and provide completeness, whereas a single level (the microstate) cannot, providing a framework where multiscaling is essential for the system's computability.

To this end, the distortion module introduced in Chap. 4 quantifies the deformation of computational structures triggered by larger scales to explore novel trajectories and make computational systems decidable [19].

For example, in cancer treatment, we find a situation where incompleteness is relevant too: "*The oxygen-deprived cells (environment) suffer an excess of DNA methylation, which silences the expression of tumor-suppressing genes, thereby enabling aberrant cellular behavior and enhancing tumor growth*"[3] [75].

This is also related to the way cancer is treated: while several efforts have focused on the identification of biological mechanisms for the targeted treatment of the disease, practical application has shown that this strategy often does not work but is harmful to the patient because cancer cells are capable of adapting quickly, causing tolerance and mutation during treatment [59].

After all, not only in molecular biology but also in ecosystems, we find living systems that continuously expand and adapt over several scales. Rivers are not just streams of water but complex systems that connect multiple systems across space, height, and even time. Even water evaporation from large river systems such as those in the Amazon creates invisible celestial rivers several hundred meters above the rai forest.

Similarly, mountains are not only bedded rocks but also complex interconnected systems with complex interdependence from the summit to the base of the mountain. They play an important role in determining global and regional climates; are the source for most rivers; act as cradles, barriers and bridges for species; and are critical to the survival and sustainability of many human societies and civilizations [58].

---

[3] http://www.genengnews.com/gen-news-highlights/cancers-grow-by-throwing-epigenetic-smo ther-parties/81253107/.

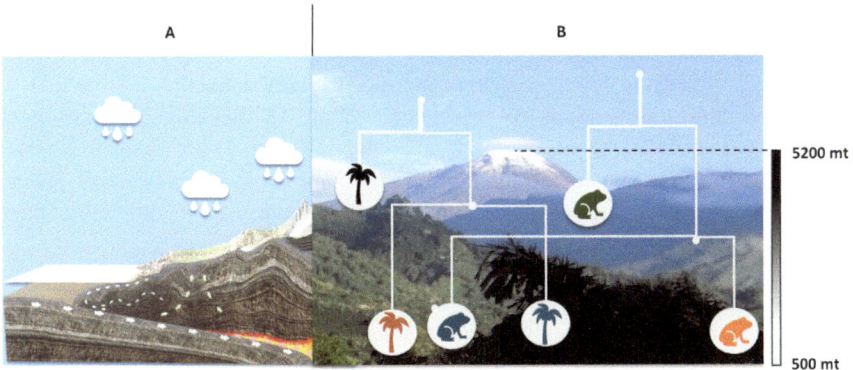

Fig. 6.3 Both topography and geology are relevant for the generation of appropriate conditions for diverse ecosystems. The drawing indicates the hidden geological and climatic diversity and processes, including orogenic rain, weathering, soil formation, sedimentation, unconformities, a fold and thrust belt and fault lines integrated into the visible aspects of a mountain system (B): biodiversity, including the phylogenetic relationships among species illustrating speciation through allopatric species (for example, frogs, which often have non-overlapping ranges), and edaphic adaptations (for example, palms, which are often confined to particular soil types such as clay or sand). The 'visible' biological and geological aspects of the mountain, including vegetational zonation, eroded landscapes and rocky outcrops, often contribute to diversification and adaptive radiations by increasing biome shifts, thus promoting biotic interactions [58]. Vulcano, Nevado del Tolima, central cordillera, Colombia. Foto/Marta Villanueva

Perhaps the fundamental understanding of life requires not only Petri dishes for experimentation but also mountain systems (see Fig. 6.3).

In general, multiple scales across several levels in complex systems allow the definition of internal reservoirs and degrees of freedom at different levels, including interacting elements or agents [64], which tend to blur the clear definition of the identities of interacting elements and their corresponding causal properties.

This means that these systems continuously define their own coarse graining [40] and, at the same time, are constantly trying to define their context (environment).

6.2.2 Elastic States and Context Selection

On the basis of this assumption, we adopted a more pragmatic approach. We used this fundamental assumption to introduce novel methods for inductive and deductive modeling.

From this last perspective, we introduce a novel complexity measurement, the complexity index Φ_K^S. The purpose of this is to evaluate the degree of variability and punctual system autonomy.

The concept of elastic states arises from the assumption that interacting elements are ultimately observers, deciding on the local context regarding its reference scale.

This has deep implications since we assume that the universe is deeply incomplete. In addition, interacting elements continuously decide their identity regarding their context, which implies that the context cannot be trivially defined.

Thus, in multiscale systems, there is no clear context (context means a family of concatenated space $\ldots \subseteq \mathcal{R}_0 \subseteq \mathcal{R}_1 \subseteq \ldots$ s, D2 in Chap. 4) that leads to a novel fundamental proposition for the foundations of complex systems in a case where an undefined context implies an incomplete local axiom, i.e., incomplete control functions.

Proposition 1 *Interacting elements in multiscale systems* (a family of concatenated spaces $\ldots \subseteq \mathcal{R}_0 \subseteq \mathcal{R}_1 \subseteq \ldots$) *with internal degrees of freedom have basic intrinsic decision-making capabilities that allow them to change and adapt their identity accordingly to their context (environment) to generate local completeness.*

Corollary 1 *The context of complex multiscale systems is not necessarily available.*

While the standard theory of complex systems implicitly assumes a notion of completeness (such as the objective definition of causal properties represented by a control function), in the present theory, we propose that completeness produced by the system cannot be trivially deconstructed from it (see Fig. 6.4).

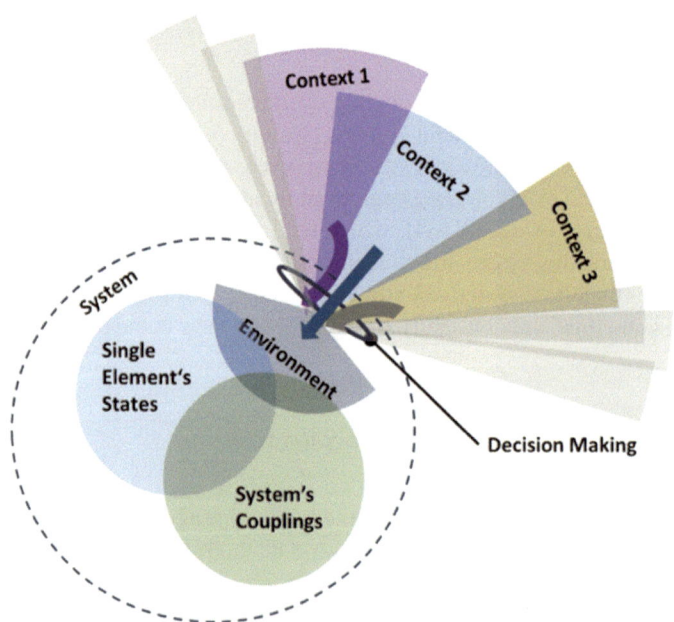

Fig. 6.4 A system built from single elements and corresponding states and coupling. The environment (context) is also essential for the full definition of a system. Here, we assume that the environment is not known a priori and that decision-making determines the system's environment depending on the specific context according to the reference scale

Let us then compare the implications of this proposition with respect to physics. Physics relies on the definition of Hamilton functions, which possess topological characteristics and even symmetries that can be optimized.

These functions are causal properties that contain notions of symmetry governing the system's state in phase space [37]. Symmetry and parsimony are concepts derived from the Noether theorem [29] used in physics to define a kind of beauty from mathematics as an intuitive guide to identify fundamental natural laws [74].

However, in our previous proposition, we argued that systems cannot be trivially reduced, nor do they lead to causal properties with universal symmetry. This is because such systems are open, have multiple scales, and cannot be characterized simply by isolating their interacting elements. According to the example presented in Chap. 4, the context of chemotactic organisms is provided by their predators; in more complex ecological systems, different populations can be considered as a context of a single organism.

Thus, it is clear that there is much more than just the difference in the understanding of fundamental concepts in complex systems (e.g., biology) and in physics [46] and that the concept of beauty cannot be easily communicated, expressed or formulated in mathematical terms for any complex system.

While physics is fundamental to the understanding of biological functions that rely on physical interactions [13], it is probably not possible to comprehend complex systems in general from the perspective of physics, as there is no a priori way to define laws in a complex system.

This limits the ability of the concept of absolute control functions, which can be objectively assessed. In addition, it sets limits for the preferred descriptive scale of the complex system. This finding indicates that in open multiscale systems, the degree of coarseness cannot always be assessed.

Thus, there is an essential need for an agency to resolve local incompleteness, which has profound implications in the way causality and determinism are understood.

In principle, Proposition 1 implies that there are basic cognitive properties that cannot be reduced to physical principles. This thesis is non-monistic and can therefore be very problematic from a conventional perspective that aims to reduce any phenomena to physical principles.

This step is rather pragmatic, as it represents an alternative way to solve several problems in complex systems (whose authors such as Christoph Koch had also recognized) [55].

Moreover, this thesis is problematic because it relies on a kind of dualism that accepts that cognition cannot be trivially reduced to physical concepts (such a discussion can be found in the concept of the Mind and Body problem [61]).

We insist that the persistence of imperfection in the world requires a pragmatic solution that cannot be reduced to physical aspects and, by adopting a nonmonistic approach, is not an act of cowardice and capitulation (by abandoning the idea of deriving self-contained theories) but rather an act of humility (embracing a concept that has also been recognized by other non-Western and nonindustrial cultures).

6.2.3 Strategies for Developing a Novel Theoretical Approach to Complex Systems

To better understand complex systems, the following steps are key:

The first step is to take a conservative approach and accept that the universe may be incomplete and therefore irreducible to fundamental axioms. A universe in which multiple scales are aggregated but cannot be fully defined requires principles that guide coordination across these scales. This step widely the accepted concepts norms with the assumption that complex systems and systems theories can be reduced to fundamental principles.

To achieve at least some local completeness, we must use fundamental decision principles as the fundamental principle (**second step**). Under this assumption, decision making is not a byproduct but rather a fundamental principle required to define an appropriate context (and coarse-grained degree) of the interacting elements. Thus:

Proposition 2 *In nonisolated systems, abstract definitions of very simple interacting elements (such as particles with specific states) are insufficient; therefore, in general, it is not possible to make an absolute objective description of these interacting elements.*

Of course, it makes sense to perform abstractions to reduce unnecessary complexity of the fundamental interacting elements (small world) and thus derive, for example, a systems theory in which different couplings allow the discovery of self-regulating functions and emergent properties.[4]

Validation efforts in well-designed experiments can validate such abstractions. However, reducing complexity and coarsening from the viewer's perspective is unfair. The relevant details may be overlooked and not captured under controlled experimental conditions.

Since decision-making is closely linked to life, one can conclude under the preceding assumption that perhaps life is not just a byproduct or an emergent phenomenon but is necessary for the creation of a coherent context.

Two deep assumptions are as follows:

First, in principle, any complex system can be dissected, which can lead to understanding from each perspective and the intent as to who is performing the dissection.

The second assumption involves understanding the entire system and how the individual elements and the system produce local completeness, which cannot always be dissected.

The first assumption implies that a dissection can always be performed and that "small worlds" can be discovered (e.g., the deconstruction of regulatory mechanisms in cells). Such knowledge can help us understand and even control complex systems in a small space (and in isolated states, according to the OACEM principle introduced in Chap. 2).

---

[4] Which is the result of this self-regulation.

Considering our initial metaphor about complex systems as labyrinths, a well-characterized "small world" system leads to a single labyrinth that allows the recognition and extraction of a notion of causality, either as causal pathways or as causal properties that allow the definition of a function projecting initial and known data or parameters into an expected output, $f : I \rightarrow O$.

However, in general, systems cannot be simply dissected, and the "*system's function*" (which is different from the mathematical function f) cannot be trivially reduced in physical terms simply because context cannot be assessed in a trivial way (see, for example, the problem of context in complex systems governance [63]).

Thus, instead of a potential representation f, there is a family of many possible representations that cannot be represented in an objective way (this implies that f tends to become a more complex object).

Thus, decomposing the system does not lead to a sustainable understanding of an open and complex system in real life. In addition, the second assumption implies that complex systems have fundamental and inherent decision-making.

Although we have proposed mathematical and theoretical arguments in support of this theory, we remain cautious. If these concepts can be verified, there are natural limits to upscaling and extrapolating systems when starting with simple microelements.

6.3 Life and Decision-Making

From the perspective of Darwinian evolution theory, life has evolved from simple principles, where the constant optimization of certain characteristics, such as optimal reproduction or dominance over other species, has led to natural selection and subsequent inheritance of advantageous phenotypes that increase the prevalence of survival and reproduction in individual organisms [43].

The theory of evolution, together with Darwin's diagram, describes species existence and extinction and inspires the concept of the fitness function. This has opened the door to mathematical modeling that describes global and local resource consumption [5].

The identification of control functions, such as fitness functions, and the constant advancement of theories of complex networks have transformed biology from a qualitative science to a more quantitative and exact science.

At the core of such development, there are simple principles, such as the central dogma of biology [12], i.e., genes as fundamental principles that code for the assembly of proteins, which serve only to define initial principles for understanding the structure, dynamics and evolution of biological systems.

This kind of genetic determinism has become the basis for understanding life, inspiring concepts such as the selfish gene and thus the radical insistence that the digital information in a gene is effectively immortal and must be the primary unit of selection.

No other entity shows such persistence—no chromosomes, no individuals, no groups and no species [60].

According to the perspective presented in the previous section, life has a completely different meaning and function than the standard dogma of biology. As D. Noble noted, nature is much more complex than a deterministic principle, and life expresses itself in many different ways, using a variety of mechanisms on multiple levels [54].

Information transmission is not limited to genes [33] but extends to different scales, such as epigenomics, diet and even the microbiome [16].

The combination of several elements, the construction of different scales and the preservation of the net-like pattern in an autopoietic way led to the development of concepts in which consciousness and pattern preservation are closely linked; such a concept was developed by Maturana and Varela and is currently being promoted at the so-called Santiago School (founded by Maturana and Varela in Santiago de Chile), which essentially aims to recognize the deep connection between cognition and biology [10, 47, 48].

In this chapter, we propose a more fundamental role for cognition and life that goes beyond the constitution of living organisms. Considering that life is connected to decision-making processes, life appears to be a unifying force in an ocean of incompleteness.

Therefore, from this perspective, life itself is much more quintessential and not limited to organisms capable of metabolizing or regulating complex processes. We are thus setting a condition in which developmental freedom plays a fundamental role[5] too [24], in which complex systems cannot be trivially reduced to single deterministic principles.

This can guide new criteria needed for the characterization of living organisms. For example, rainforests or river deltas are often considered ecosystems of geographical importance. Since organisms are joined together in an ecosystem, creating different structures on different scales and defining and redefining their context, these systems behave just like whole organisms.

This can have profound implications for how an organism is defined, as this concept is not necessarily limited to cells, animals, plants, or fungi [10, 82] but rather applies to all aspects of any kind of system embedded at several scales, capable to own internal reservoirs and requiring a continuous definition of their context.[6]

Thus, life is perhaps more than the constant and self-sustained regulation of energy and information flows, constructing the building blocks of complex systems (organisms) with units (organisms) interacting with the environment and organized at different scales [10]. The constant decision-making process and the implicit ability to decide in different contexts, resulting in the final definition of the organism's environment, implies that the concept of life can be more ubiquitous [40] and less restricted by any rigid boundary defining a unit (see Fig. 6.5).

---

[5] https://plato.stanford.edu/entries/theories-biological-development/.

[6] Such definition could have also an impact in the way how extraterrestrial life is defined and identified.

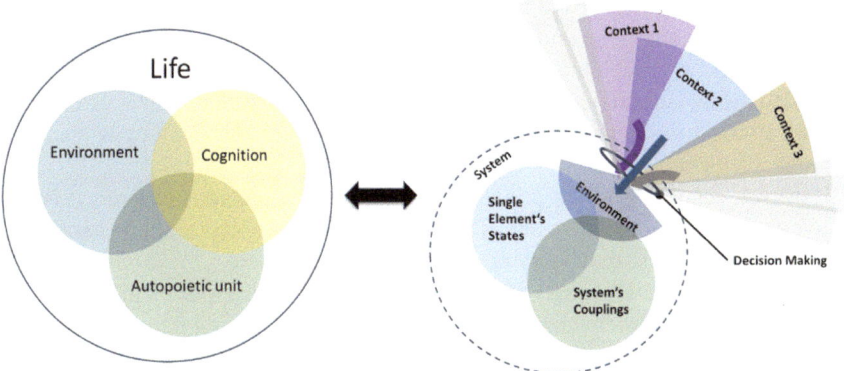

Fig. 6.5 Usually, life is studied considering closed unities; even holistic approaches, such as the Santiago School, consider the combination of an autopoietic unity, environment perception and cognition into a single closed unity. Instead, we consider that systems in general are open and incomplete regarding different possible contexts. We observe that we do not refer to "life" since we consider that these processes are fundamental and that the qualities characterizing "life" are ubiquitous

In this way, life has a purpose and is **probably rooted in cognitive processes continuously seeking an agreement to find a common context** ($\{\ldots \subseteq \mathcal{R}_0 \subseteq \mathcal{R}_1 \subseteq \ldots\} \rightarrow \mathcal{R}_i$, i.e., reduce one context from a family of concatenated contexts), **thus reducing persistent incompleteness arising from multiscale systems**.

Furthermore, and following a similar thesis proposed by Koch [40], cognitive processes are probably not limited to only organic systems, which could imply a new definition of life (a definition more related to natural and intrinsic decision processes and not restricted to only biological systems).

We are cautious about introducing these concepts and believe that more research is needed. However, defining life in this way offers an entirely new thought-provoking perspective in the natural sciences that both challenges accepted ways of making and understanding systems and, at the same time, offers new perspectives on basic ethical definitions.

6.3.1 Consequences in Control Theory

In the previous sections, we discussed the implications of an incomplete universe, where different contexts must be decided upon and are not definable a priori. In such a universe, decision making is continuously needed.

On the basis of this argument, we have hypothesized that life can be something more fundamental and not simply an emergent effect. In this section, we examine the possible consequences of this assumption for control theory.

Control requires knowledge to predict and control any system. Are we able to control every system in the universe, from small quantum systems to large galaxies that cross cells? Although this span over several scales is absurd, it is an example of man's hedonistic ambition to control every corner of the universe. This is often done initially at small scales and then systematically by scaling larger systems made up of simple elements.

However, according to the principle proposed in the previous chapter, it is not always possible to assess in an objective way the causal properties of a system. This presents a challenge in the theory of control.

In the previous sections, we have also discussed the implications of an incomplete universe in which different contexts must be decided and are not definable a priori. A universe such as ours requires continuous decision-making.

On the basis of this argument, we hypothesized that life can be something more fundamental and not just an emergent effect. In this section, we examine the possible consequences of this assumption for control theory.

6.3.1.1   Limitations of Control Theory

Addressing persistent uncertainty, in which elements take on elastic states in response to their context, poses a challenge to conventional systemic approaches, which isolate complex systems into simple, well-defined interacting entities (the small-world approach).

The challenge lies in the fact that such systematic isolation removes the context of an element. It also ignores how systems define themselves by their own coarse grain processes when interacting with other systems.

If the elements have some kind of necessary fundamental agency, then they are agents that continuously evaluate and decide the most appropriate possible states depending on their context. Moreover, this poses a problem for any objective explanation of a system, since it cannot be ignored that elements are also agents and of itself that choose to reduce the inherent incompleteness of their context.

This also implies that there is a problem in modern complex science: the insistence on the concept of "self" without context (e.g., self-organization), as if linkage and coupling were indeed the fundamental basis for defining the identity and thus the function of a system, conceptually blurs the dependence of the elements and the entire system in context.

Thus, the context is continuously created by the system. In addition, other external systems also create their own context.

Despite the apparent success of several studies, they address only specific questions on specific time, space, and size scales. This poses a problem for control theory (as a philosophical challenge) and control theory (in a branch of engineering).

For both theories, a single agent who acts intentionally and has a mental state extracts causal relationships, which is causal potency, to exert control over a system [68].

6.3.1.2 Shared Control with Other Interacting Elements

From this perspective, only agents with advanced cognitive characteristics can possess this type of causal potency and agency. Nonetheless, they often ignore the fact that other elements are also agents making their own decisions.

For example, hydroelectric power plants have been designed and built by considering hydrodynamics, geology and the interaction of various forces in the project structure. However, such designs often ignore biology, ecology and the entire environment, with very large dams being disrupted and even irreparably destroying entire ecosystems [9].

Despite attempts to restore geological and ecological conditions, such as the massive reinforcement of mountains to avoid landslides or close monitoring and even the reintroduction of fish species into the Yang Tse River after the construction of the Three Gorges Dam, all these efforts have had little or no success, as damage has appeared to be irreparable [35, 72].

In addition, international trade has created conditions for the reproduction of invasive species that disrupt some of these large-scale projects. The zebra mussel, for example, is an invasive species that has established itself in North America under very good conditions and has spread throughout all tributaries and seas. Owing to their strong feet, they can be easily attached to any surface, including pipelines that transport water under high pressure, blocking the flow of water through these pipelines and disrupting the operation of large hydroelectric power plants[7] [31].

6.3.1.3 Limits of Upscaling Control from the "Small World" to the "Large World"

Therefore, while in engineering, evaluating the impact of projects and systems on ecosystems is challenging [77], sooner or later, different actors at different levels could find an opportunity and use such engineering projects to propagate and colonize new niches, perhaps disrupting the original large engineering project.

Restricting the dismantling of a system implies that technology will inevitably lead to disruption rather than control.

The extrapolation and upscaling of small worlds have profound consequences; for example, a small dam in a river system may not be able to profoundly disrupt an ecosystem and can be implemented sustainably.

Nevertheless, a large-scale project inevitably destroys large systems but also creates dependencies that prevent a reversal to its original state.

This is perhaps why the massive dam projects in Kazakhstan, which dried up the Aral Sea, led to one of the most terrible dystopias, which changed and irreparably damaged the entire ecosystem and is one of the best examples of an ecosystem

---

[7] https://www.waterpowermagazine.com/analysis/controlling-invasive-mussels/.

collapse [38]. Similar examples can also be found all over the world with similar plight of disruptions.[8]

The same can also be said about how the world has globalized, with a world based solely on the extraction of promoting economic and material growth in every region of the globe.

Recognizing these boundaries, limitations and impacts is essential not only for understanding nature but also for modeling engineering projects, creating societies, and defining economics.

Control theory and economics are therefore very relevant and closely linked. Controlling a system is highly valuable. Since the economy is highly dependent on technical and new technological advances, there is a general tendency to create high expectations for high returns on investment by putting pressure on the advancement of technological solutions that show promising results in the laboratory and often ignoring potential risks in the introduction of such technologies on a large scale.

Recognizing potential risks and scaling limits can prevent the use of technologies that deliver a significant return on investment for a few investors in a short period of time. However, these technologies are very expensive and have irreversible long-term effects.

6.3.2 Consequences in Artificial Intelligence

The study of human intelligence naturally leads to the problem of defining cognition and consciousness. To this end, a clear definition is needed to understand the degree of depth gained from such research.

The "soft" problem refers to cognition as a product of brain processes, whereas the "hard" problem refers to personal experiences [10, 11].

Currently, several projects focus on soft problems to better understand hard problems. For example, the Flagship Project "The European Human Brain" aims to understand this problem by exploring and simulating cerebral processes.[9]

Additionally, conventional research on artificial intelligence focuses on the exploration of synthetic systems replicating brain processes to imitate human capabilities [8], including not only classical neural networks but also mechanisms such as attention to extracting data (and thus information) from a context [80].

Interestingly, machine learning models, particularly large language models (LLMs), are currently used as paradigms to understand the connection between brain function and consciousness [44] and are thus employed to address the difficult problem of consciousness (despite the fact that these models are essentially the representation—and not the real modelling—of a consciousness space, compressing

---

[8] Unfortunately, large projects are a way to signalize prestige and power and are often promoted by authoritarian regimes to signalize their power.

[9] https://www.humanbrainproject.eu/en/.

all the potential predictions obtained from cognitive patterns extracted from different data sources [49]).

Life, cognition, and multiscaling are so deeply interconnected that cognitive abilities do not simply emerge from brain activity alone and are fundamental in navigating the environment. This perspective has been suggested by different authors, such as R. Penrose, who suggested that consciousness cannot be represented algorithmically, suggesting that it is the product of quantum mechanical processes in the mind [56], and Luisi, who also suggested that consciousness is deeply connected to the process of life itself [10].

Furthermore, several investigations have recently recognized the role of the gut and its gut microbiota in brain activity and its potential influence on cognition [15]. Such research therefore raises the question of whether consciousness is the effect of cerebral activity or whether consciousness is also connected to the whole body (mind–body problem).

Finally, some authors acknowledge that consciousness can even be associated with systems that are different from the brain, in which the mere connectivity of interlocking objects can lead to fundamental consciousness "atoms" [40].

The theory presented in this book is nonalgorithmic and does not distinguish between soft and hard problems. Instead, we follow the notion of "cognitive atoms", which are deeply linked to living processes (as proposed by the Santiago School [47]); according to this idea, a basal cognition is fundamental to the 'gluing' of complex systems since we assume that consciousness is deeply related to the system's environment to resolve its context (and reduce persistent incompleteness in the system).

From this perspective, consciousness cannot be isolated from a context (see Thesis 1 in this chapter). Furthermore, according to this notion, consciousness is fundamental and not simply a phenomenon or "object" that can be created.

Thus, multiple systems connected across multiple scales may be responsible for generating a type of cognition (similar to the relationship between the gut microbiota and the brain [15]).

Current research agendas on artificial intelligence (AI) involve the transformation of data into information. Of course, conventional AI research is not free from the natural human tendency to attribute intelligence and cognition to physical objects. However, if the hypothesis as a factor gluing different scales is correct, then cognition cannot be simply simulated or modeled by mimicking cerebral states (see Fig. 6.6, with a graphical summary and workflow of the theories of consciousness starting from the concept of the "small world", as discussed in this section).

Therefore, the question is not how cognition is objectively measured from models and synthetic brain-like systems ("phenomenal consciousness"), but what are people's intuitions about conscious phenomena in artificial intelligence [14].

Finally, current research agendas are too human-centric [7], considering consciousness as being something that can be found only in humans, ignoring the fact that we also have technical and conceptual limitations in accessing the feeling in and of itself in other organisms and/or even recognizing the cognitive capabilities of plants [50].

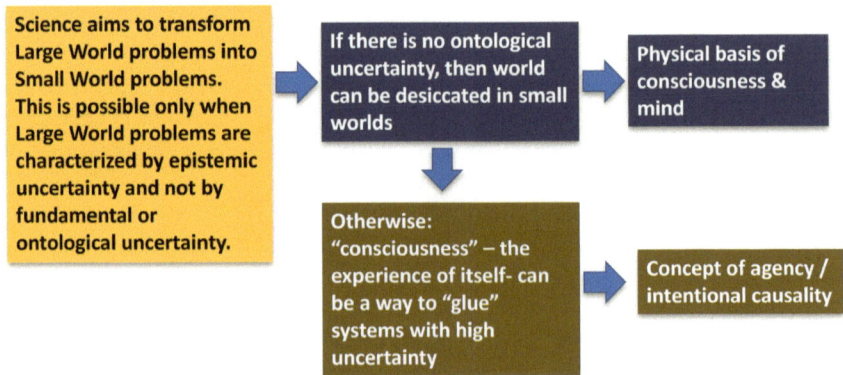

Fig. 6.6 Diagram visualizing different possible pathways to understand the problem of consciousness depending on how complex systems can be dissected into small worlds [81]

Therefore, the clues for a better understanding of the consciousness and experience of oneself probably do not lie in big data, such as large models, and eventual emergent properties but rather in accepting the fact that the agency and the experience of oneself are also ways to reduce natural and persistent incompleteness, thus guaranteeing the existence of interconnected elements across multiple scales. It is therefore impossible to escape from a labyrinth within a labyrinth without consciousness.

6.4 From Control to Communication: Semiotics

All theoretical approaches to complex systems, especially control theory, can be limited in that they reduce any interaction to physical terms and ignore the possibility that the interactions of a system are directly related to its function and importance. It is therefore obvious that communication in complex systems is more important than the mere exchange of information.

Reaction pathways and protein interactions in cells, for example, have been identified as a type of signaling in complex systems. However, signaling, as a form of communication, is perhaps much more than a simple metaphor (as Capra and Luisi noted [10]).

A complex system confined by physical constraints evolves into emergent, self-sustaining processes that function with only a small number of species that possess very simple properties. Watson and Lovelock originally proposed the daisy world model, in which two types of organisms on a planet orbiting the sun can maintain a population and produce homeostasis with variable light intensity [82].

Such a model has been proposed to support Lovelock's Gaia theory and describes how homeostasis arises. Despite the fact that both computer simulations and experiments reproduce the original conditions under which life merged (e.g., the famous experiments of Stanley Miller in 1953 [34]), observing how complex signaling pathways arise in cells, as well as how phenotypes such as colors and different forms emerge in biological systems, is still striking.

Nature is more than a simple energy balance and exchange of information. How this information is interpreted, and its purpose considering the interpretation of a given context, can take many forms.

According to Proposition 1, communication can only be relevant if basic decision-making and cognitive skills are required to determine the context of the system and its identity to interact in a given context.

Communication is not limited to the exchange of information, such as encoding or decoding instructions. To determine the context of the exchanged information or even a signal, the message must be associated with a form of communication that conveys meaning.

If this hypothesis is correct, the diversity of phenotypes and the inclusion of different sensory tools are crucial when making decisions in complex environments characterized by different contexts.

This could have profound implications for how we understand interactions in complex systems. Complex systems are often disrupted to derive internal mechanisms. However, if communication and meaning are relevant, then the reduction of communication to simple signals and the flow of substances (e.g., in cell signaling) could mean that a crucial element for the correct understanding of complex systems is still missing.

Some scientists have recently claimed that understanding natural systems requires the explicit inclusion of intrinsic cognitive aspects, such as an observer's perspective, in the system's study (which, of course, requires some form of communication) and that this knowledge is relevant to reversing the problems that technology and development have created in the past.

These approaches include recognizing indigenous knowledge of nature and the environment [39] and understanding the role that emotions play in regard to understanding nature [83].

In spite of this, indigenous perceptions are often dismissed [18]. However, such concepts, in which animism is a central theme, have been relevant in numerous cultures over time and are used today as valuable sources of information [85]. It has been demonstrated that such attitudes offer a more sustainable way of life than modern industrialized and globalized Western cultures do[10] [39, 62] and provide valuable information that will help implement techniques such as permaculture.

The recognition of meaning in communication, beyond the mere study of interactions in complex systems, is a field that, at least in biological systems, belongs to the concept of biosemiotics.[11]

This requires the interpretation of signs not only as information but also in relation to the context in which they are being exchanged (for example, considering the role

---

[10] https://theconversation.com/animism-recognizes-how-animals-places-and-plants-have-power-over-humans-and-its-finding-renewed-interest-around-the-world-181389.

[11] https://www.biosemiotics.org/what-is/.

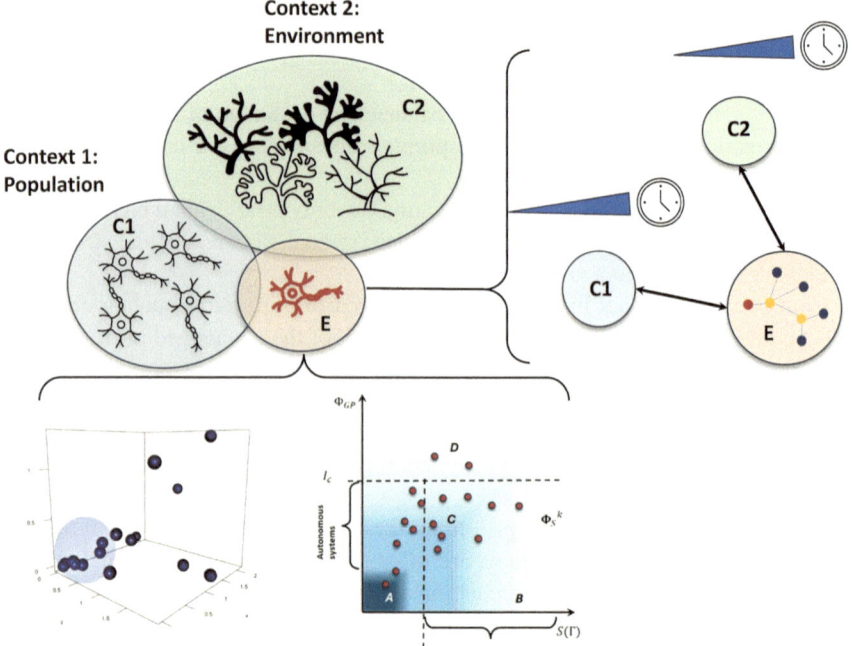

Fig. 6.7 Accordingly, when they are viewed in context (different concatenated spaces ... $\subseteq \mathcal{R}_0 \subseteq \mathcal{R}_1 \subseteq$ ... mean, in this example, to define either the population where the element belongs (C1) or the whole environment considering other organisms (C2)), they influence and must be explicitly incorporated into the models of the system. The method can be considered contextual learning, akin to the methods recently developed in language modeling [80]

of communication in the construction of communities and nonhuman cultures[12] [32], such as the combination of codas used by sperm whales depending on the context [67]).

Signal transmission is crucial, but interference makes the system vulnerable, making it easy to hack. For example, parasites such as fungi can control ants and program them to dip their mandibles into the succulent main vein of a leaf.

As soon as the ant dies, a mushroom stem grows from its head. When the stem grows, the fungus sheds its spores [17]. However, signals propagate not only information but also meaning and that should be understood as a form of communication.

The fact that signaling and communication are so central implies that systems require more than holistic descriptions. Contextual learning, i.e., the analysis of data patterns related to their context (see Fig. 6.7), can be an approach to this problem.

Recently, transformer-based NLP models have attracted much attention in biology and genetics because of their ability to process variable-length input sequences in parallel and use self-attention mechanisms to capture long-term dependencies [41].

---

[12] Including the complex.

The attention mechanism, i.e., the consideration of context in modeling [80], together with the training of large amounts of unlabeled data represents a paradigm shift in the way models are constructed: with attention, the goal is to train models to resect a given context, which makes them much more universal and flexible (for example, by combining attention with graphs [66], as well as by using cross training on different datasets [7]).

By using such models, called foundation models, it is possible to define rather homogeneous models that can then be further trained to accomplish specific tasks [7].

By constructing "holistic graphs", where singular observables associated with the observed entity (element E in Fig. 6.7) can be connected to contexts 1 and 2 with different degrees of coarse graining in a single graph ($C1$ and $C2$ in Fig. 6.7), it is then possible to use methods, like attention mechanism, applied to graphs and combined with graph neural networks [73] to extract patterns for observables associated with E while considering the different contexts, assuming that E is communicating with $C1$ and $C2$ (see Fig. 6.7).

An easy way to extract possible communication and semiotics in complex systems is to combine the estimation of entropy (evaluation of the degree of autonomy and variability of the system) with the potential integration of information from the system and its environment, taking into account different degrees of coarse grains and using attention-graph learning [73] (see Fig. 6.7).

Here, it is important to realize that human language (as a form of communication) and semiotics should be used to recognize all possible contexts that are relevant to make a model for calculating predictions complete, since not all contexts are known a priori (in contrast to functional models—like LLM models—that sample all the potential and available contexts to compute predictions of word sequences—which ignores structural aspects such as grammar—[49]).

Thus, language probably does not function as a mechanism but rather as a means of navigating through persistently unknown environments.

In this context, language as a form of communication is also perhaps not exclusive to humans. Here, we aim to avoid anthropomorphism (assume that other organisms share the same, or similar, qualities with humans) as well as anthropectomy (i.e., assume that other organisms do not share certain qualities with humans) and follow a line of thinking proposed by Berwick and Chomsky, who said *"It's about as likely that an ape will prove to have a language ability as there is an island somewhere with a species of flightless birds waiting for humans to teach them to fly"* [4].

To consider communication, and even language, goes beyond pure data analysis and machine learning techniques: it is about the importance of the transmitted signal sequences, i.e., the analysis of the signal, the environment, and the intention of the receiver and the manufacturer. The observables are therefore connected not only to the system of interest but also to its surroundings.

This implies that there is no way to break down the system into simple parts and then upscale it by increasing its complexity (small-world approach). Instead, the observation should consider the system and its context (big world), which is a challenging task because the boundaries between the system and its context are not always easy to identify.

Furthermore, since there is no a priori way to access the communication between the system and its environment, long-term observations are needed (which extends beyond compressing information via, for example, long short-term memory (LSTM) methods to compress time series [71]).

As a result of such long-term observations, it might take more than one generation to truly understand a system; this might suggest that there is a kind of "slow science" where the expected results take many years or even centuries to arrive.

6.5 Implications for Technology, Economics and Ethics

Humanity is currently facing dramatic tribulations ever experience in human history.[13] It is caused by a series of events with global impact, such as climate change, pandemics, social and cognitive distortion in citizens on social media networks, and geo-terrorism.[14]

Unfortunately, all these dangers are in part due to dramatic advances in science and technology (starting from the atomic bomb to the development of cognitive agents and artificial intelligence). We live in an era of ultrafast scientific and technological development, which has its origin in the infinite possibilities that the combination of technologies and principles provides to generate novel discoveries and technologies.[15]

This enabling combination of modules led to an exponential growth of innovations according to Moore's law [51].[16]

Our lifestyles have changed as a result of this technological development, partly owing to our own coevolution with technology, with an insatiable appetite for new innovative products, and a naïve credo of science and technology as a unique way to improve our lives has evolved over time.

At the same time, humans have developed large structures and technologies, such as cities or production complexes, that behave like large organisms that are disconnected from other ecosystems.

Technology is power. This explains why it is common for any technology to find its way first for military rather than civilian purposes; this fact, the weapon of every scientific discovery and technology (including technologies such as social networks

---

[13] In the year when this essay was written: 2024.

[14] Inflicted by a few individuals and dictators of power disrupting societies. By practicing corruption, they are able to impose dominance by brute force on already shattered societies and to take the world hostage by using weapons of mass destruction to achieve their goals and consolidate their power [79].

[15] https://www.bbc.com/future/article/20190207-technology-in-deep-time-how-it-evolves-alongside-us.

[16] https://www.cs.utexas.edu/~fussell/courses/cs352h/papers/moore.pdf.

or AI), reflects the high risk and danger we pose from our science, which in this way makes science essentially futile, if not dangerous for our own survival as a species.

Finally, could technology have its own needs? Considering that cognitive abilities are likely to be fundamental and not newly created, there is no chance that any technology will start acting autonomously. Therefore, there is no man-made technology that also grows and evolves naturally.

Technologies do not belong to an environment per se. This is especially true when such technologies are made from synthetic materials and processes. Consequently, technologies should be maintained by humans to ensure their proper functioning. This causes many types of technologies to behave like organisms that are decoupled from the environment but have essential functions (including metabolism) maintained by human workers (such as factories).

In addition, people are seeing and perceiving the world and environment through their technologies rather than going out there connecting with other human beings, which means that societies are becoming increasingly disconnected from the environment and each other.

Therefore, we must finally realize that the narrative of science and technology as benefactors of humanity may be an illusion after observing the current state of the world. This rapid creation of innovative discoveries and several potential discoveries (from biology to physics) have side effects that could worsen our situation and further endanger our existence, as symbiotic technology—the human ends up as a closed, system-accelerated system that ignores long-lasting processes and their relationships within the context.

As we discussed in the subsection on control theory, once these novel technologies enter the market (such as CRISP-CAR or quantum computation), there will be new challenges and dangers, as these technologies will be an integral part of economies, production systems and societies. Since the markets do not follow moral principles, it will be too late to stop all the negative effects of these technologies once they are standardized.

Because of this disillusionment with technological progress, there should be a smart barrier between basic research and technology transfer. With the help of this barrier, we can intelligently slow the process of technology transfer while assessing its potential negative consequences when developing technologies from short to long term without hindering the appropriate advancement of technology and the potential actual benefits of these new technologies for people, society as a whole and the environment in the long run (consider, for instance, the petition to stop training LLMs before their societal impact is well understood[17]).

For example, as part of the approval process for new drugs, agencies such as the Food and Drug Administration (FDA) and the European Medicine Administration (EMA) are currently assessing technological risks. However, such organisms only collect evidence of direct adverse effects and do not evaluate effects on different time and space scales, such as feeding data [53].

---

[17] https://futureoflife.org/open-letter/pause-giant-ai-experiments/.

Furthermore, they cannot create incentives to slow progress to obtain solid evidence about the long-term effects of these novel technologies or design clear boundary conditions for their broader use.

We need to learn about our communication and relation with the environment before setting up a technology.[18] For example, machines and chemical technologies can be combined in agriculture to increase the yield of large soja monocultures, ignoring biodiversity and the relationship between the soja and the environment, or alternative small cultures can combine soja with other plants and techniques to preserve local biodiversity and good "communication" between species—but perhaps with a lower yield (for example, using permaculture [10]).

This takes time, and it is therefore necessary to accept that technologies cannot be trivially scaled up (and that they can be efficiently applied only locally and not on a global scale). In addition, we must also be aware that there is probably no way to control a system.

Consequently, it is essential to think in terms of **deep time** and innovations that allow us to communicate with our environment instead of creating artificial and self-consuming organisms. Our development as a society should be based on constant negotiation (and not control) with other organisms and our environment.

6.6 Outlook

In this thesis, we have provided methods for identifying persistent entropy as well as methods for representing complex networks consisting of interacting elements with unstable identities. While the first approach is useful for characterizing an observed system, the second approach is more suitable for building models and studying the consequences of system shifts according to the local perspective.

Our ability to observe and model each system therefore depends on both the autonomy and variability of this complexity. To this end, we derived a method to measure the internal bias of system mechanisms as they continuously adapt to their context [19].

In addition, we derived a second variable based on persistent entropy to quantify the distortion of system structures compared with other systems (system variability) as well as autonomous adaptation to new contexts (system autonomy) [20].

This information is then encoded in a vector-like complexity measure, called Φ_S entropy, to measure such persistent entropy and is recognized when we deal with a simple and observable labyrinth (small persistent entropy) or when there is a labyrinth in a labyrinth (large persistent entropy).

---

[18] Perhaps for this reason, it is reasonable to think that our civilization, society and technology are an accident. This development is not necessarily a deterministic pathway across the universe. Even though other forms of intelligence can develop more advanced forms of communication with their environment, due to their nature they remain unnoticed. Furthermore, the application of techniques developed to investigate extraterrestrial civilizations could be used to make contact with other species on earth.

Thus, the Φ_S entropy is a useful method for identifying causal principles or effects and thus constructing system theories.

The first consequence of this assumption is to find methods to measure the influence of different perspectives on the observability and modeling of complex systems. Beyond that, there are likely limits to a single objective perspective. However, to this end, the evaluation of Φ_S must be much more efficient.

The second consequence is the effect that different perspectives on natural systems can have on our understanding of science and technology. If different perspectives on nature play a role, then causal potency and human ingenuity should change to make up more perspectives than the purely human perspective does (Fig. 6.4).

Real control does not mean imagining a control function to model a system, which ultimately serves only the goals of a few individuals belonging to a selected few in society. Instead, a real understanding of complex systems and their control should also consider other perspectives, such as the individual perspective of organisms of their environment or the role of signals and communication across several scales.

The emulation of a multifunction shared by multiple agents in complex systems is not only a way to respect different perspectives in complex systems but also a path to more sustainable societies. Thus, the use of the Φ_S complexity should be included in decision making and the design of future scenarios.

The Φ_S entropy can be considered a recognition of boundaries where system theory and control theory can be applied or where an alternative approach is needed.

A possible alternative described in this book is improving communication with other systems, a concept that is only sketched here (see the interpretation of the Φ_S diagram in Fig. 6.8, where the concept considered in Fig. 6.4 is also included in the interpretation).

Much remains to be done to better characterize and understand how multiscaling arises and how elements within such multiscaling decide and communicate. In this case, semiotics, combined with theoretical computer science and machine learning, as well the active inclusion of indigenous knowledge and the active recognition of the conscious inner workings of other living beings [28], could be relevant to understanding signs, codes and communication in complex systems [25].

In this sense, we must seek communication with our own environment more than communication with extraterrestrial life.[19]

Given the impact of coevolution between society and technology, the idea of implementing and/or improving our communication with our environment could mean going beyond the concept of *collapsology*, which involves analyzing various factors accounting for social and environmental tipping points before the systems collapse [65]—a concept that requires systems to be observed objectively, which is not always the case, as we analyzed in Chap. 5—and going deeper into the communication web across complex systems.

---

[19] Currently, all the knowledge from SETI research could be perhaps used to improve the communication with our own world and fellow species.

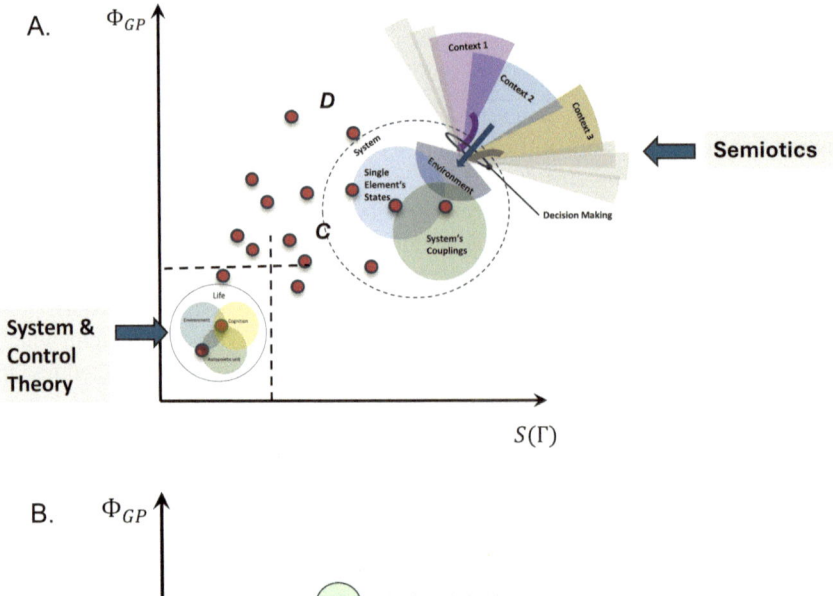

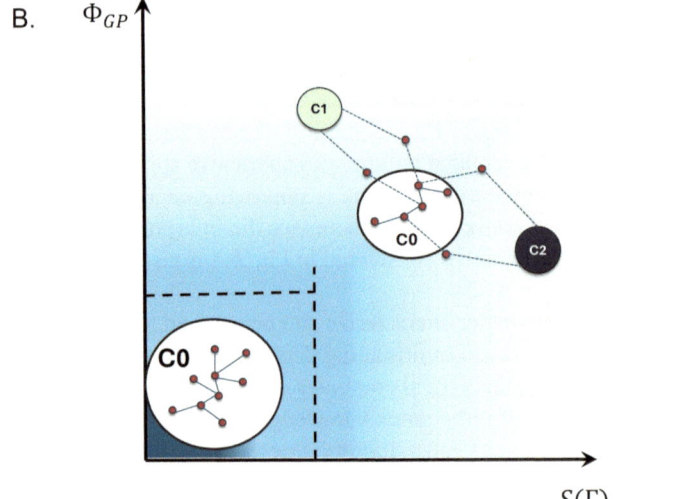

Fig. 6.8 a Representation of modeling boundaries according to the Φ_S diagram and combined with Fig. 6.4. In the lower left corner, standard methods for control theory and systems theory can be identified. As a result of the perspective presented in this chapter, the systems identified on the open right side cannot be trivially reduced to systems theory; instead, a better understanding of all possible contexts (as concatenated spaces $\ldots \subseteq \mathcal{R}_0 \subseteq \mathcal{R}_1 \subseteq \ldots$) as well as communication involving signs and codes is necessary in such cases (semiotics). **b** For example, nodes with well-defined interactions can be represented in a network that provides the context for individual nodes (C0); if more than one context is relevant, individual nodes cannot be trivially defined as a connection to the original network, and their relationship (and communication) to additional contexts should also be considered

6.7 Concluding Remarks

To systematically understand complex systems, we have argued that the concept of small worlds is essential. Following the metaphor introduced at the beginning of this book, this approach aims to understand the structure of a labyrinth. This approach is comparable to trying to escape a labyrinth, in which, using basic input parameters or datasets I, we compute a given prediction O via a well-defined mathematical function or operator $\hat{f}$ that guides us through the labyrinth, particularly $\hat{f} : I \rightarrow O$.

However, we have also discussed persistent entropy and elastic states and their possible role in achieving some degree of incompleteness in complex systems, which is in the form of a maze, and address the problem of persistent entropy, thus implying that the mathematical function $\hat{f}$ the mathematical model of the system cannot be derived in an objective way, making $\hat{f}$ a much more complex mathematical object.

The small-world method and concepts of causality (presented in Chap. 3) led us to the Icarus tragedy: the genius of Daedalus allowed him to build a labyrinth so skillfully that once it was built, he was unable to escape it himself; however, after these creatures escaped from the labyrinth, Daedalus and his son were imprisoned by King Minos in his own labyrinth so that he started making wings using feathers, thread, and beeswax for both Icarus, his son and him to escape the labyrinth by flight (or by flying out of the Labyrinth).

After Icarus had flown out, he disobeyed his father by flying to the sky; however, the sun melted the wax of his wings (which held the feathers together), and he fell down. When Icarus flapped his wings, he looked like a bird. He then realized that he had no more feathers and began to flutter his featherless arms. The loose feathers and warm, sticky wax started dripping off Icarus's arms. Every feather started to fall like snowflakes and down, down, and down until it plummeted into the sea (where it drowned).

This story could serve as an analogy in explaining why, despite science and technology and a growing awareness of current problems and possible solutions, there is no real willingness to act and/or address these problems. In a way, we are like Icarus fleeing from the labyrinth only to fall to the sea.

Technically, individuals and society are cognitively overwhelmed by all the available knowledge. However, a lack of skills in addressing this knowledge and the current challenges posed by complex systems and their apparent control might lead to detrimental assumptions and fatal decisions.[20]

The thesis presented in this chapter is that cognitive abilities are essential in deciding on interacting elements at different scales and in achieving consensus to minimize persistent incompleteness in the world and perhaps in the entire universe.

Thus, cognition and information processing are perhaps a consequence but the fundamental principle when multiscale systems emerge, in which the minimization of persistent incompleteness and the reduction of chained contexts to a single context

---

[20] There is a danger of dehumanization (including loss of empathy) through technology and information overload. Here is important to remember the Rusell–Einstein Manifesto: "Remember your humanity and forget the rest".

can or must be considered a fundamental purpose associated with living systems (such a fundamental principle can perhaps be described in what is currently defined in biology as basal cognition becomes [45]).

In this context, we purport that life, and all its interconnected forms of functioning, are capable of continuously generating completeness from imperfection in a way that cannot be trivially reduced to physical processes or seen only as a byproduct of emergent processes. As a result, life cannot be seen as a consequence but rather as something fundamental.

According to what we discussed in Chap. 2, this comes together with the 1st and 4th forms of Aristotelian causality, recognizing that there is not only a physical but also a functional causality that cannot always be accessed objectively.

Together with this hypothesis, there are not only consequences for the fundamental understanding of the foundations of the world and the universe. In addition, there are consequences in the way we do science and translate science into technology.

Therefore, we must strive to find ways to communicate even better both within our own species and between other species, to be aware of warning signs, and to recognize the limits of our knowledge and technological efforts [3]; technology can be the answer in this case if it is used not only to control boundaries but also to detect boundaries and when we evolve and change with our technologies in such a way that we are better off coupling (and not decoupled) with our environment.

As the philosopher Hannah Arendt put it in 1958, in recent centuries, we have developed a science "that considers the nature of the Earth from the viewpoint of the Universe"[21] [2].

However, in doing so, we have paradoxically trained ourselves to ignore the most critical lesson of all: our coevolution with technology[22] [2]. Science and technology separate us from nature. They create an illusion of absolute controllability and observability of a complex system. In spite of this, it is this same science and technology that could help us overcome barriers to understanding and communicating with our environment.

Science and technology should promote different kinds of development and knowledge that are deeply integrated into the complex, especially in biological systems, on the basis of their ability to communicate and cooperate with other organisms.

Currently, we are attempting to combine our current technological advances and knowledge, by using them with humility, to better understand other species and complex systems. In providing such tools for understanding and communication—and even language across nature [4]—we hope to reinforce the underlying message that science and technology should be based on humility rather than greed by recognizing the fundamental value of life that is close to what L. Wittgenstein claimed [84]:

Die Welt und das Leben sind eins

---

[21] https://www.bbc.com/future/article/20190207-technology-in-deep-time-how-it-evolves-alongside-us.

[22] https://iep.utm.edu/hannah-arendt/#H4.

(The world and life as one)

Domaso (Italy)/Stuttgart (Germany)/Rif. Rivetti (Italy)—May–July–August 2024

References

1. Ali A (2023) Biodiversity–ecosystem functioning research: brief history, major trends and perspectives. Biol Conserv 285:110210. https://doi.org/10.1016/j.biocon.2023.110210
2. Arendt H, Canovan M (1998) The human condition, 2nd edn. University of Chicago Press, Chicago
3. Barnosky AD, Hadly EA, Bascompte J, Berlow EL, Brown JH, Fortelius M, Getz WM, Harte J, Hastings A, Marquet PA, Martinez ND, Mooers A, Roopnarine P, Vermeij G, Williams JW, Gillespie R, Kitzes J, Marshall C, Matzke N, Mindell DP, Revilla E, Smith AB (2012) Approaching a state shift in earth's biosphere. Nature 486:52–58. https://doi.org/10.1038/nature11018
4. Berwick RC, Chomsky N (2017) Why only us: language and evolution, Reprint. The MIT Press, Cambridge, Massachusetts, London, England
5. Bessonov N, Reinberg N, Volpert V (2014) Mathematics of Darwin's diagram. Math Model Nat Phenom 9:5–25. https://doi.org/10.1051/mmnp/20149302
6. Bitbol M, Luisi PL (2004) Autopoiesis with or without cognition: defining life at its edge. J R Soc Interface 1:99–107. https://doi.org/10.1098/rsif.2004.0012
7. Bommasani R, Hudson DA, Adeli E, Altman R, Arora S, von Arx S, Bernstein MS, Bohg J, Bosselut A, Brunskill E, Brynjolfsson E, Buch S, Card D, Castellon R, Chatterji N, Chen A, Creel K, Davis JQ, Demszky D, Donahue C, Doumbouya M, Durmus E, Ermon S, Etchemendy J, Ethayarajh K, Fei-Fei L, Finn C, Gale T, Gillespie L, Goel K, Goodman N, Grossman S, Guha N, Hashimoto T, Henderson P, Hewitt J, Ho DE, Hong J, Hsu K, Huang J, Icard T, Jain S, Jurafsky D, Kalluri P, Karamcheti S, Keeling G, Khani F, Khattab O, Koh PW, Krass M, Krishna R, Kuditipudi R, Kumar A, Ladhak F, Lee M, Lee T, Leskovec J, Levent I, Li XL, Li X, Ma T, Malik A, Manning CD, Mirchandani S, Mitchell E, Munyikwa Z, Nair S, Narayan A, Narayanan D, Newman B, Nie A, Niebles JC, Nilforoshan H, Nyarko J, Ogut G, Orr L, Papadimitriou I, Park JS, Piech C, Portelance E, Potts C, Raghunathan A, Reich R, Ren H, Rong F, Roohani Y, Ruiz C, Ryan J, Ré C, Sadigh D, Sagawa S, Santhanam K, Shih A, Srinivasan K, Tamkin A, Taori R, Thomas AW, Tramèr F, Wang RE, Wang W, Wu B, Wu J, Wu Y, Xie SM, Yasunaga M, You J, Zaharia M, Zhang M, Zhang T, Zhang X, Zhang Y, Zheng L, Zhou K, Liang P (2022) On the opportunities and risks of foundation models. https://doi.org/10.48550/arXiv.2108.07258
8. Bringsjord S, Govindarajulu NS (2024) Artificial intelligence. In: Zalta EN, Nodelman U (eds) The Stanford encyclopedia of philosophy. Metaphysics Research Lab, Stanford University
9. Briones-Hidrovo A, Uche J, Martínez-Gracia A (2021) Hydropower and environmental sustainability: a holistic assessment using multiple biophysical indicators. Ecol Indic 127:107748. https://doi.org/10.1016/j.ecolind.2021.107748
10. Capra F, Luisi PL (2014) The systems view of life: a unifying vision, 1st edn. Cambridge University Press
11. Chalmers DJ (2010) The character of consciousness. Oxford University Press. https://doi.org/10.1093/acprof:oso/9780195311105.001.0001
12. Cobb M (2017) 60 years ago, Francis Crick changed the logic of biology. PLoS Biol 15:e2003243. https://doi.org/10.1371/journal.pbio.2003243
13. Cockell CS (2017) The laws of life. Phys Today 70:42–48. https://doi.org/10.1063/PT.3.3493
14. Colombatto C, Fleming SM (2024) Folk psychological attributions of consciousness to large language models. Neurosci Conscious 2024:niae013. https://doi.org/10.1093/nc/niae013

15. Cooke MB, Catchlove S, Tooley KL (2022) Examining the influence of the human gut micro-biota on cognition and stress: a systematic review of the literature. Nutrients 14:4623. https://doi.org/10.3390/nu14214623

16. Costa dos Santos G, Renovato-Martins M, de Brito NM (2021) The remodel of the "central dogma": a metabolomics interaction perspective. Metabolomics 17:48. https://doi.org/10.1007/s11306-021-01800-8

17. de Bekker C, Ohm RA, Loreto RG, Sebastian A, Albert I, Merrow M, Brachmann A, Hughes DP (2015) Gene expression during zombie ant biting behavior reflects the complexity underlying fungal parasitic behavioral manipulation. BMC Genomics 16:1–23. https://doi.org/10.1186/s12864-015-1812-x

18. De Smedt J, De Cruz H (2023) Animisms: practical indigenous philosophies. In: Smith T (ed) Animism and philosophy of religion. Springer International Publishing, Cham, pp 95–122. https://doi.org/10.1007/978-3-030-94170-3_5

19. Diaz Ochoa JG (2018) Elastic multi-scale mechanisms: computation and biological evolution. J Mol Evol 86:47–57. https://doi.org/10.1007/s00239-017-9823-7

20. Diaz Ochoa JG (2023) A unified method for assessing the observability of dynamic complex systems. Comput Biol Med 160:107012. https://doi.org/10.1016/j.compbiomed.2023.107012

21. Diaz Ochoa JG, Bucher J, Pery AR, Zaldivar Comenges JM, Niklas J, Mauch K (2013) A multi-scale modeling framework for individualized, spatiotemporal prediction of drug effects and toxicological risk. Front Pharmacol 3. https://doi.org/10.3389/fphar.2012.00204

22. Donovan C (2019) Biological function. In: Shackelford TK, Weekes-Shackelford VA (eds) Encyclopedia of evolutionary psychological science. Springer International Publishing, Cham, pp 1–4. https://doi.org/10.1007/978-3-319-16999-6_2097-1

23. Eisenstein M (2024) Foundation models build on ChatGPT tech to learn the fundamental language of biology. Nat Biotechnol 42:1323–1325. https://doi.org/10.1038/s41587-024-02400-2

24. Fagan MB, Maienschein J (2022) Theories of biological development. In: Zalta EN (ed) The Stanford encyclopedia of philosophy. Metaphysics Research Lab, Stanford University

25. Favareau D (2009) Essential readings in biosemiotics, biosemiotics. Springer Netherlands, Dordrecht. https://doi.org/10.1007/978-1-4020-9650-1

26. Gardner M (1970) Mathematical games [WWW document]. Sci Am. URL https://www.scientificamerican.com/article/mathematical-games-1970-10/. Accessed 24 May 2024

27. Gaskill M (n.d.) Rise in roadkill requires new solutions [WWW document]. Sci Am. URL https://www.scientificamerican.com/article/roadkill-endangers-endangered-wildlife/. Accessed 27 Aug 2024

28. Godfrey-Smith P (2017) Other minds: the octopus, the sea, and the deep origins of consciousness, Reprint. Farrar, Straus and Giroux, New York

29. Goldstein H, Poole C, Safko J (2001) Classical mechanics, 3rd edn. Pearson, San Francisco, Munich

30. Haken H (2012) Synergetics: an introduction, 3rd edn. Softcover reprint of the original 3rd edn. 1983 edition. Springer

31. Haubrock PJ, Soto I, Kourantidou M, Ahmed DA, Serhan Tarkan A, Balzani P, Bego K, Kouba A, Aksu S, Briski E, Sylvester F, De Santis V, Archambaud-Suard G, Bonada N, Cañedo-Argüelles M, Csabai Z, Datry T, Floury M, Fruget J-F, Jones JI, Lizee M-H, Maire A, Murphy JF, Ozolins D, Jessen Rasmussen J, Skuja A, Várbíró G, Verdonschot P, Verdonschot RCM, Wiberg-Larsen P, Cuthbert RN (2024) Understanding the complex dynamics of zebra mussel invasions over several decades in European rivers: drivers, impacts and predictions. Oikos 2024:e10283. https://doi.org/10.1111/oik.10283

32. Hersh TA, Gero S, Rendell L, Cantor M, Weilgart L, Amano M, Dawson SM, Slooten E, Johnson CM, Kerr I, Payne R, Rogan A, Antunes R, Andrews O, Ferguson EL, Hom-Weaver CA, Norris TF, Barkley YM, Merkens KP, Oleson EM, Doniol-Valcroze T, Pilkington JF, Gordon J, Fernandes M, Guerra M, Hickmott L, Whitehead H (2022) Evidence from sperm whale clans of symbolic marking in non-human cultures. Proc Natl Acad Sci U S A 119:e2201692119. https://doi.org/10.1073/pnas.2201692119

33. Hewitt SM (2020) Negative consequences of the central dogma. J Histochem Cytochem 68:731. https://doi.org/10.1369/0022155420970927
34. Hill HGM, Nuth JA (2003) The catalytic potential of cosmic dust: implications for prebiotic chemistry in the solar nebula and other protoplanetary systems. Astrobiology 3:291–304. https://doi.org/10.1089/153110703769016389
35. Hvistendahl M (n.d.) China's three gorges dam: an environmental catastrophe? [WWW document]. Sci Am. URL https://www.scientificamerican.com/article/chinas-three-gorges-dam-disaster/. Accessed 25 May 2024
36. Jabr F (2019) How beauty is making scientists rethink evolution. New York Times
37. Jensen HJ (2023) Complexity science: the study of emergence, New. Cambridge University Press, Cambridge, United Kingdom, New York, NY
38. Keith DA, Rodríguez JP, Rodríguez-Clark KM, Nicholson E, Aapala K, Alonso A, Asmussen M, Bachman S, Basset A, Barrow EG, Benson JS, Bishop MJ, Bonifacio R, Brooks TM, Burgman MA, Comer P, Comín FA, Essl F, Faber-Langendoen D, Fairweather PG, Holdaway RJ, Jennings M, Kingsford RT, Lester RE, Nally RM, McCarthy MA, Moat J, Oliveira-Miranda MA, Pisanu P, Poulin B, Regan TJ, Riecken U, Spalding MD, Zambrano-Martínez S (2013) Scientific foundations for an IUCN red list of ecosystems. PLoS ONE 8:e62111. https://doi.org/10.1371/journal.pone.0062111
39. Kimmerer RW (2015) Braiding sweetgrass: indigenous wisdom, scientific knowledge and the teachings of plants. Milkweed Editions, Minneapolis, MN
40. Koch C (2019) The feeling of life itself: why consciousness is widespread but can't be computed. The MIT Press. https://doi.org/10.7551/mitpress/11705.001.0001
41. Le NQK (2023) Leveraging transformers-based language models in proteome bioinformatics. Proteomics 23:e2300011. https://doi.org/10.1002/pmic.202300011
42. Lenharo M (2023) If AI becomes conscious: here's how researchers will know. Nature. https://doi.org/10.1038/d41586-023-02684-5
43. Lennox J, Pence CH (2024) Darwinism. In: Zalta EN, Nodelman U (eds) The Stanford encyclopedia of philosophy. Metaphysics Research Lab, Stanford University
44. Li J, Li J (2024) Memory, consciousness and large language model. https://doi.org/10.48550/arXiv.2401.02509
45. Lyon P, Keijzer F, Arendt D, Levin M (2021) Reframing cognition: getting down to biological basics. Philos Trans R Soc B Biol Sci 376:20190750. https://doi.org/10.1098/rstb.2019.0750
46. MacArthur BD (2021) Truth and beauty in physics and biology. Nat Phys 17:149–151. https://doi.org/10.1038/s41567-020-01132-9
47. Maturana HR, Varela FJ (1980) Autopoiesis and cognition: the realization of the living, 1980th edn. Springer, Dordrecht, Holland, Boston
48. Maturana HR, Varela FJ (2004) El Arbol del conocimiento: las bases biológicas del entendimiento humano. Lumen, Santiago de Chile, Buenos Aires
49. Millière R, Buckner C (2024) A philosophical introduction to language models—part I: continuity with classic debates. https://doi.org/10.48550/arXiv.2401.03910
50. Montgomery BL (2024) Smarty plants? Controversial plant-intelligence studies explored in new book. Nature 629:280–281. https://doi.org/10.1038/d41586-024-01275-2
51. Moore GE (1998) Cramming more components onto integrated circuits. Proc IEEE 86:82–85. https://doi.org/10.1109/JPROC.1998.658762
52. Mukhopadhyay R, Nath S, Kumar D, Sahana N, Mandal S (2024) Basics of the molecular biology: from genes to its function. In: Anjoy P, Kumar K, Chandra G, Gaikwad K (eds) Genomics data analysis for crop improvement. Springer Nature, Singapore, pp 343–374. https://doi.org/10.1007/978-981-99-6913-5_14
53. Neltner TG, Alger HM, Leonard JF, Maffini MV (2013) Data gaps in toxicity testing of chemicals allowed in food in the United States. Reprod Toxicol 42:85–94. https://doi.org/10.1016/j.reprotox.2013.07.023
54. Noble D (2006) The music of life: biology beyond the genome. Oxford University Press. https://doi.org/10.1093/oso/9780199295739.001.0001
55. O'Gieblyn M (2019) Do we have minds of our own? New Yorker

56. Penrose R (1995) Shadows of the mind: a search for the missing science of consciousness, Neuauflage. Vintage, London
57. Penrose R (2007) The road to reality: a complete guide to the laws of the universe. Reprint Edition. Vintage, New York
58. Perrigo A, Hoorn C, Antonelli A (2020) Why mountains matter for biodiversity. J Biogeogr 47:315–325. https://doi.org/10.1111/jbi.13731
59. Prasad V (2016) Perspective: the precision-oncology illusion. Nature 537:S63–S63. https://doi.org/10.1038/537S63a
60. Ridley M (2016) In retrospect: the selfish gene. Nature 529:462–463. https://doi.org/10.1038/529462a
61. Robinson H (2023) Dualism. In: Zalta EN, Nodelman U (eds) The Stanford encyclopedia of philosophy. Metaphysics Research Lab, Stanford University
62. Rout M, Reid J (2020) Embracing indigenous metaphors: a new/old way of thinking about sustainability. Sustain Sci 15:945–954. https://doi.org/10.1007/s11625-020-00783-0
63. Schoenberg M (2022) Context in complex systems governance. In: Keating CB, Katina PF, Chesterman Jr CW, Pyne JC (eds) Complex system governance: theory and practice. Springer International Publishing, Cham, pp 209–240. https://doi.org/10.1007/978-3-030-93852-9_8
64. Schweitzer F (2007) Multi-agent approach to the self-organization of networks. https://doi.org/10.48550/arXiv.0704.2533
65. Servigne P, Stevens R (2020) How everything can collapse: a manual for our times, 1st edn. Polity, Cambridge, UK, Medford, MA
66. Shang J, Ma T, Xiao C, Sun J (2019) Pre-training of graph augmented transformers for medication recommendation. https://doi.org/10.48550/arXiv.1906.00346
67. Sharma P, Gero S, Payne R, Gruber DF, Rus D, Torralba A, Andreas J (2024) Contextual and combinatorial structure in sperm whale vocalisations. Nat Commun 15:3617. https://doi.org/10.1038/s41467-024-47221-8
68. Shepherd J (2014) The contours of control. Philos Stud 170:395–411. https://doi.org/10.1007/s11098-013-0236-1
69. Siegenfeld AF, Bar-Yam Y (2020) An introduction to complex systems science and its applications. Complexity 2020:6105872. https://doi.org/10.1155/2020/6105872
70. Silk J (2015) Physics: the impulse of beauty. Nature 523:156–157. https://doi.org/10.1038/523156a
71. Staudemeyer RC, Morris ER (2019) Understanding LSTM—a tutorial into long short-term memory recurrent neural networks. https://doi.org/10.48550/arXiv.1909.09586
72. Stone R (2008) Three gorges dam: into the unknown. Science 321:628–632. https://doi.org/10.1126/science.321.5889.628
73. Sun C, Li C, Lin X, Zheng T, Meng F, Rui X, Wang Z (2023) Attention-based graph neural networks: a survey. Artif Intell Rev 56:2263–2310. https://doi.org/10.1007/s10462-023-10577-2
74. The character of physical law [WWW document] (n.d.) MIT Press. URL https://mitpress.mit.edu/9780262560030/the-character-of-physical-law/. Accessed 25 May 2024
75. Thienpont B, Steinbacher J, Zhao H, D'Anna F, Kuchnio A, Ploumakis A, Ghesquière B, Van Dyck L, Boeckx B, Schoonjans L, Hermans E, Amant F, Kristensen VN, Koh KP, Mazzone M, Coleman ML, Carell T, Carmeliet P, Lambrechts D (2016) Tumour hypoxia causes DNA hypermethylation by reducing TET activity. Nature 537:63–68. https://doi.org/10.1038/nature19081
76. Tolman RC (1979) The principles of statistical mechanics, New edn. Dover Publications Inc., New York, NY
77. Vaccari DA, Strom PF, Alleman JE (2005) Environmental biology for engineers and scientists, 1st edn. Wiley-Interscience, Hoboken, NJ
78. Vaidyanathan B, Haraburda B, Jacobi CJ (2023) Beauty in biology: an empirical assessment. J Biosci 48:15. https://doi.org/10.1007/s12038-023-00342-6

79. van der Wal Z (2021) Singapore's corrupt practices investigations bureau: guardian of public integrity. In: Boin A, Fahy LA, 't Hart P (eds) Guardians of public value: how public organisations become and remain institutions. Springer International Publishing, Cham, pp 63–86. https://doi.org/10.1007/978-3-030-51701-4_3

80. Vaswani A, Shazeer N, Parmar N, Uszkoreit J, Jones L, Gomez AN, Kaiser L, Polosukhin I (2023) Attention is all you need. https://doi.org/10.48550/arXiv.1706.03762

81. Viale R (2021) The epistemic uncertainty of COVID-19: failures and successes of heuristics in clinical decision-making. Mind Soc 20:149–154. https://doi.org/10.1007/s11299-020-00262-0

82. Watson AJ, Lovelock JE (1983) Biological homeostasis of the global environment: the parable of Daisyworld. Tellus B 35B:284–289. https://doi.org/10.1111/j.1600-0889.1983.tb00031.x

83. Weber A (2016) Biology of wonder: aliveness, feeling and the metamorphosis of science. New Society Publishers, Gabriola Island

84. Wittgenstein L, Schulte J (1999) Tractatus logico-philosophicus, 10th edn. Suhrkamp, Frankfurt am Main

85. Zidny R, Sjöström J, Eilks I (2020) A multi-perspective reflection on how indigenous knowledge and related ideas can improve science education for sustainability. Sci Educ 29:145–185. https://doi.org/10.1007/s11191-019-00100-x

86. Zwaenepoel A, Roovers P, Hermy M (2006) Motor vehicles as vectors of plant species from road verges in a suburban environment. Basic Appl Ecol 7:83–93. https://doi.org/10.1016/j.baae.2005.04.003

Index

A
Anthropectomy, 143
Aristoteles, 14

B
Basal decision making, 5
Bayes, 16, 18
Betti number, 101

C
Canonical ensemble, 68, 122
Causal determinism, 4, 12–15, 21, 24, 25,
    51, 59
Causal inference, 3, 4, 29, 52, 63, 64, 92,
    109, 113, 121
Causality, 2–4, 8–16, 19, 20, 23–25, 29, 30,
    34, 43–45, 50, 51, 53–55, 91, 93,
    105, 108, 122, 124, 131, 133, 149,
    150
Cell, 2, 5, 16, 19, 21, 22, 30, 46, 54, 59, 64,
    73, 76, 82–84, 86, 87, 92, 128, 132,
    134, 136, 140, 141
Cell signaling, 5, 141
Coarse graining, 36, 45, 110, 129, 143
Coevolution, 39, 86, 144, 147, 150
Collapsology, 147
Completeness, 2, 13, 19, 22, 70–72, 87,
    128, 130, 132, 150
Complex network, 4, 5, 21, 23, 29, 40,
    51–54, 58, 59, 74, 92, 94, 116, 133,
    146
Consciousness, 4, 24, 43, 107, 125, 126,
    134, 138–140
Context (in multiscales), 70

Control function, 15, 31–34, 36–40, 42, 43,
    48, 51, 52, 57, 74–77, 79, 122, 124,
    128, 130, 131, 133, 147
Control theory, 8, 29, 135, 136, 138, 140,
    145, 147, 148
Critical state, 23, 24, 34, 35, 58, 124

D
Deductive modelling, 4, 19, 20, 30, 31, 35,
    42, 47, 51, 55, 129
Deep time, 146
Detailed balance, 33
Determinism, 4, 12, 14, 16, 24, 42, 64, 124,
    131, 133
Differential topology, 54
Digitalism, 43
Distortion matrix, 102, 103
DNA, 21, 55, 56, 128
Doctrine of causes (Aristotelian), 10, 14

E
Eco, Umberto, 1
Elastic states, 4, 5, 63, 71, 74, 88, 91 94,
    97, 118, 129, 136, 149
Electronic health records, 59
Element (interacting), 5, 7, 18, 23, 30–33,
    35, 36, 38, 40, 42, 43, 47, 52, 54, 57,
    58, 63–71, 73–75, 78–81, 86, 87, 91,
    105, 107, 121, 122, 124, 126–132,
    137, 146, 149
Emergence, 4, 23, 34, 42, 43, 71, 88, 124
Energy (function), 37
Ensemble, 32, 44, 68, 74, 122

© The Editor(s) (if applicable) and The Author(s), under exclusive license
to Springer Nature Switzerland AG 2025
J. G. Diaz Ochoa, *Complexity Measurements and Causation for Dynamic Complex Systems*,
Understanding Complex Systems, https://doi.org/10.1007/978-3-031-84709-7